U.S. Department
of Transportation

Federal Aviation
Administration

AVIATION MECHANIC GENERAL

Practical Test Standards

June 2003

Supplied by:

JEPPESEN SANDERSON, INC.
55 Inverness Drive East
Englewood, CO 80112-5498

FLIGHT STANDARDS SERVICE
Washington, DC 20591

NOTE

FAA-S-8081-26, Aviation Mechanic General Practical Test Standards (PTS) is to replace the oral and practical test guides currently used. Both testing procedures will be in effect until all examiners have been trained to administer the test in accordance with the PTS, or 2 years after the effective date of Order 8610.4J, Aviation Mechanic Examiner Handbook. After which time, **all** tests must be administered under the PTS guidelines. New examiners must use the PTS upon completion of initial training. Previously appointed examiners must transition to the PTS within 60 days after completion of recurrent training.

Record of Changes

Change 1: 8/8/2003

Introduction

Performance Levels
    LEVEL 1—Z3b. Nondestructive changed to specified.
    LEVEL 1—PERFORMANCE STANDARD deleted.
    LEVEL 2—bullet 2: added additional text.
    LEVEL 2—PERFORMANCE STANDARD deleted.
    LEVEL 3—bullet 4: added additional text.
    LEVEL 3—Z3e. Verify changed to check.
    LEVEL 3—PERFORMANCE STANDARD deleted.

Change 2: 9/24/2003

Introduction

Performance Levels
    LEVEL 1—PERFORMANCE STANDARD added.
    LEVEL 2—PERFORMANCE STANDARD added.
    LEVEL 3—PERFORMANCE STANDARD added.

Section 1—Aviation Mechanic General

A. Basic Electricity, Objective 1. Change "at least four" to "at least two."
B. Aircraft Drawings, Objective 1. Change "at least four" to "at least two."
C. Weight and Balance, Objective 1. Change "at least four" to "at least two."
D. Fluid Lines and Fittings, Objective 1. Change "at least four" to "at least two."
D. Fluid Lines and Fittings, Objective 2 a. Added the words "fabrication and tubefittings.
E. Materials and Processes, Objective 1. Change "at least four" to "at least two."
F. Ground Operation and Servicing, Objective 1. Change "at least four" to "at least two."
G. Cleaning and Corrosion Control, Objective 1. Change "at least four" to "at least two."
H. Mathematics, Objective 1. Change "at least four" to "at least two."
I. Maintenance Forms and Records, Objective 1. Change "at least four" to "at least two."
J. Basic Physics, Objective 1. Change "at least four" to "at least two."
K. Maintenance Publications, Objective 1. Change "at least four" to "at least two."
L. Aviation Mechanic Privileges and Limitations, Objective 1. Change "at least four" to "at least two."

Change 3: 6/21/2004

Introduction

Unsatisfactory Performance (Page 6)

Section 1—Aviation Mechanic General

A. Basic Electricity
B. Aircraft Drawings
C. Weight and Balance
E. Materials and Processes
F. Ground Operation and Servicing
G. Cleaning and Corrosion Control
K. Maintenance Publications

FOREWORD

This Aviation Mechanic General Practical Test Standards book has been published by the Federal Aviation Administration (FAA) to establish the standards for the Aviation Mechanic General Practical Test. The passing of this practical test is a required step toward obtaining the Aviation Mechanic certificate with Airframe and/or Powerplant ratings. **FAA inspectors and Designated Mechanic Examiners (DMEs) shall conduct practical tests in compliance with these standards.** Applicants should find these standards helpful in practical test preparation.

/s/ 2-13-2003

Joseph K. Tintera, Manager
Regulatory Support Division
Flight Standards Service

CONTENTS

Introduction.. 1

    Practical Test Standard Concept 2
    Practical Test Book Description 2
    Practical Test Standard Description 2
    Use of the Practical Test Standards 3
    Aviation Mechanic Practical Test Prerequisites 4
    Examiner Responsibility 4
    Performance Levels 4
    Satisfactory Performance 6
    Unsatisfactory Performance 6

SUBJECT AREAS

SECTION I—AVIATION MECHANIC GENERAL

  A. BASIC ELECTRICITY 1-1
  B. AIRCRAFT DRAWINGS 1-2
  C. WEIGHT AND BALANCE 1-3
  D. FLUID LINES AND FITTINGS 1-4
  E. MATERIALS AND PROCESSES 1-5
  F. GROUND OPERATION AND SERVICING 1-6
  G. CLEANING AND CORROSION CONTROL 1-7
  H. MATHEMATICS 1-8
  I. MAINTENANCE FORMS AND RECORDS 1-9
  J. BASIC PHYSICS 1-10
  K. MAINTENANCE PUBLICATIONS 1-11
  L. AVIATION MECHANIC PRIVILEGES AND
      LIMITATIONS 1-12

INTRODUCTION

The Flight Standards Service of the Federal Aviation Administration (FAA) has developed this practical test book as a standard to be used by FAA inspectors and Designated Mechanic Examiners (DMEs) when conducting aviation mechanic practical tests. Applicants are expected to use this book when preparing for practical testing.

Information considered directive in nature is described in this practical test document in terms, such as "shall" and "must" indicating the actions are mandatory. Guidance information is described in terms, such as "should" and "may" indicating the actions are desirable or permissive but not mandatory.

The FAA gratefully acknowledges the valuable assistance provided by the many individuals and organizations that contributed their time and talent in assisting with the development of these practical test standards.

This practical test standard may be downloaded from the Regulatory Support Division's, AFS-600, web site at http://afs600.faa.gov. Subsequent changes to this standard, in accordance with AC 60-27, Announcement of Availability: Changes to Practical Test Standards, will also be available on AFS-600's web site and then later incorporated into a printed revision.

This publication can be purchased from the Superintendent of Documents, U.S. Government Printing Office, Washington, DC 20402. The official online bookstore web site for the U.S. Government Printing Office is www.access.gpo.gov

Comments regarding this document should be sent to:

> U.S. Department of Transportation
> Federal Aviation Administration
> Regulatory Support Division
> Airman Testing Standards Branch, AFS-630
> P.O. Box 25082
> Oklahoma City, OK 73125

FAA-S-8081-26

Practical Test Standard Concept

Title 14 of the Code of Federal Regulations (14 CFR) specifies the subject areas in which knowledge and skill must be demonstrated by the applicant before the issuance of an Aviation Mechanic Certificate with an airframe and/or powerplant rating. The CFRs provide the flexibility that permits the FAA to publish practical test standards containing knowledge and skill specifics in which competency must be demonstrated.

"Knowledge" (oral) elements are indicated by use of the words *"Exhibits knowledge of...."*

"Skill" (practical) elements are indicated by the use of the words *"Demonstrates the ability to...."*

The FAA will revise this book whenever it is determined that changes are needed. **Adherence to the applicable regulations, the policies set forth in the current revision of FAA Order 8610.4, Aviation Mechanic Examiner Handbook, and the practical test standards is mandatory for the evaluation of aviation mechanic applicants.**

Practical Test Book Description

This test book contains the following Aviation Mechanic Practical Test Standards.

Section I—Aviation Mechanic General

Practical Test Standard Description

The Aviation Mechanic Practical Test Standards include the subject areas of knowledge and skill for the issuance of an aviation mechanic certificate and/or the addition of a rating. The subject areas are the topics in which aviation mechanic applicants must have knowledge and/or demonstrate skill.

The REFERENCE identifies the publication(s) that describe(s) the subject area. Descriptions of the subject area are not included in the practical test standards, because this information can be found in references listed and/or in manufacturer or FAA-approved or acceptable data related to each subject area. Publications other than those listed may be used as references if their content conveys substantially the same information as the referenced publications. Except where appropriate, (e.g., pertinent CFRs) references listed in this document are NOT meant to supersede or otherwise replace manufacturer or other FAA-approved or acceptable data, but to serve as general information and study material sources. **Information contained in manufacturer and/ or FAA-approved/acceptable data always takes precedence over advisory or textbook referenced data.** Written instructions given to applicants for the completion of assigned skill portions of the practical test standard may include service bulletins; airworthiness directives or other CFRs; type certificate data sheets or specifications; manufacturer maintenance manuals or other similar approved/acceptable data necessary for accomplishment of objective testing.

Reference List:

14 CFR part 1	Definitions and Abbreviations
14 CFR part 21	Certification Procedures for Products and Parts
14 CFR part 43	Maintenance, Preventive Maintenance Rebuilding, and Alteration
14 CFR part 65	Certification: Airmen Other Than Flight Crewmembers
14 CFR part 91	Air Traffic and General Operating Rules
ABS	Aircraft Basic Science, Glencoe— Macmillan/McGraw-Hill Publishing Co.
AC 65-9	Airframe and Powerplant Mechanics General Handbook
AC 65-15	Airframe and Powerplant Mechanics Airframe Handbook
AEE	Aircraft Electricity and Electronics, Glencoe— Macmillan/MacGraw-Hill Publishing Co.
AMR	Aircraft Maintenance and Repair, Glencoe— Macmillan/MacGraw-Hill Publishing Co.
AMT-G	Aviation Maintenance Technician Series— General, Aviation Supplies and Academics (ASA), Inc.
FAA-H-8083-1	Aircraft Weight and Balance Handbook
JSAT	A & P Technician Airframe Textbook—Jeppesen Sandersen, Inc.
JSGT	A & P Technician General Textbook—Jeppesen Sandersen, Inc.

Each subject area has an objective. The objective lists the important knowledge and skill elements that must be utilized by the examiner in planning and administering aviation mechanic tests, and that applicants must be prepared to satisfactorily perform.

EXAMINER is used in this standard to denote either the FAA Inspector or FAA Designated Mechanic Examiner (DME) who conducts the practical test.

Use of the Practical Test Standards

The FAA requires that all practical tests be conducted in accordance with the appropriate Aviation Mechanic Practical Test Standards and the policies and procedures set forth in the current revision of FAA Order 8610.4. When using the practical test book, the examiner must evaluate the applicant's knowledge and skill in sufficient depth to determine that the objective for each subject area element selected is met.

An applicant is not permitted to know before testing begins which selections in each subject area are to be included in his/her test (except the core competency elements, which all applicants are required to perform). Therefore, an applicant should be well prepared in *all* oral and skill areas included in the practical test standard.

Further information about the requirements for conducting/taking the practical test is contained in FAA Order 8610.4

Aviation Mechanic Practical Test Prerequisites

All applicants must have met the prescribed experience requirements as stated in 14 CFR part 65, section 65.77. (See FAA Order 8610.4 for information about testing under the provisions of 14 CFR part 65, section 65.80.)

Examiner Responsibility

The examiner who conducts the practical test is responsible for determining that the applicant meets acceptable standards of knowledge and skill in the assigned subject areas within the appropriate practical test standard. Since there is no formal division between the knowledge and skill portions of the practical test, this becomes an ongoing process throughout the test.

The following terms may be reviewed with the applicant prior to, or during, element assignment.

1. "Inspect" means to examine by sight and/or touch (with or without inspection enhancing tools/equipment).
2. "Check" means to verify proper operation.
3. "Troubleshoot" means to analyze and identify malfunctions.
4. "Service" means to perform functions that assure continued operation.
5. "Repair" means to correct a defective condition.

Performance Levels

The following is a detailed description of the meaning of each level.

Level 1

- Know basic facts and principles.
- Be able to find information and follow directions and written instructions.
- Locate methods, procedures, instructions, and reference material.
- Interpretation of information not required.
- No skill demonstration is required.

Z3b.   Locate specified nondestructive testing methods. (Level 1)

Performance Standard: The applicant will locate information for nondestructive testing.

Level 2

- Know and understand principles, theories, and concepts.
- Be able to find and interpret maintenance data and information, and perform basic operations using the appropriate data, tools, and equipment.
- A high level of skill is not required.

Example:

Z3c.   Detect electrical leakage in electrical connections, terminal strips, and cable harness (at least ten will have leakage faults). (Level 2)

Performance Standard: Using appropriate maintenance data and a multimeter, the applicant will identify items with leakage faults.

Level 3

- Know, understand, and apply facts, principles, theories, and concepts.
- Understand how they relate to the total operation and maintenance of aircraft.
- Be able to make independent and accurate airworthiness judgments.
- Perform all skill operations to a return-to-service standard using appropriate data, tools, and equipment. Inspections are performed in accordance with acceptable or approved data.
- A fairly high skill level is required.

Example:

Z3e.   Check control surface travel.   (Level 3)

Performance Standard: Using type certificate data sheets and the manufacturer's service manual, the applicant will measure the control surface travel, compare the travel to the maintenance data, and determine if the travel is within limits.

Satisfactory Performance

The practical test is passed if the applicant demonstrates the prescribed proficiency in the assigned elements (core competency and other selected elements) in each subject area to the required standard. Applicants shall not be expected to memorize all mathematical formulas that may be required in the performance of various elements in this practical test standard. However, where relevant, applicants must be able to locate and apply necessary formulas to obtain correct solutions.

Unsatisfactory Performance

If the applicant does not meet the standards of any of the elements performed (knowledge, core competency, or other skill elements), the associated subject area is failed, and thus the practical test is failed. The examiner or the applicant may discontinue testing any time after the failure of a subject area. In any case, the applicant is entitled to credit for only those subject areas satisfactorily completed. See the current revision of FAA Order 8610.4 for further information about retesting and allowable credit for subject areas satisfactorily completed.

Typical areas of unsatisfactory performance and grounds for disqualification include the following.

1. Any action or lack of action by the applicant that requires corrective intervention by the examiner for reasons of safety.
2. Failure to follow acceptable or approved maintenance procedures while performing skill (practical) projects.
3. Exceeding tolerances stated in the maintenance instructions.
4. Failure to recognize improper procedures.
5. The inability to perform to a return to service standard, where applicable.
6. Inadequate knowledge in any of the subject areas.

SECTION I—AVIATION MECHANIC GENERAL

A. BASIC ELECTRICITY

*Core competency element.

REFERENCES: JSGT; AEE; AMT-G.

Objective.   To determine that the applicant:

1. Exhibits knowledge of at least two of the following—

   a. sources and/or effects of capacitance in a circuit.
   b. uses of capacitance in a circuit.
   c. sources and/or effects of inductance in a circuit.
   d. uses of inductance in a circuit.
   e. operation of basic AC and/or DC electrical circuits.
   f. Ohm's law.
   g. Kirchoff's law(s).
   h. procedures used in the measurement of voltage, current, and/or resistance.
   i. determining power used in simple circuits.
   j. troubleshooting, and/or repair or alteration using electrical circuit diagrams.
   k. common types of defects that may occur in an installed battery system.
   l. aircraft battery theory/operation.
   m. servicing aircraft batteries.

2. *Demonstrates the ability to perform both of the following—

   a. use measuring equipment to measure in a circuit or circuit component(s), at least one of the following: voltage, current, resistance, or continuity. (Level 3)
   b. determine the appropriateness of measurement(s) according to instructions/specifications. (Level 2)

3. Demonstrates the ability to perform at least one of the following—

   a. read and interpret one or more electrical circuit diagrams. (Level 2)
   b. troubleshoot an electrical circuit. (Level 3)
   c. calculate voltage, current, and resistance using Ohm's Law. (Level 2)
   d. inspect a battery and installed battery system. (Level 3)
   e. accomplish a battery state-of-charge (hydrometer) and/or electrical leak (cell imbalance) test. (Level 3)

FAA-S-8081-26

    f.   accomplish removal and/or installation of a battery in an aircraft. (Level 3)

    g.   set-up and connect a charger to one or more batteries for constant current and/or constant voltage charging. (Level 3)

B. AIRCRAFT DRAWINGS

REFERENCES: ABS; JSGT; AMT-G.

Objective.   To determine that the applicant:

1.   Exhibits knowledge of at least two of the following—

    a.   characteristics and/or uses of any of the various types of drawings/blueprints and/or system schematics.

    b.   the meaning of any of the lines and symbols commonly used in aircraft sketches/drawings/blueprints.

    c.   using charts or graphs.

    d.   troubleshooting an aircraft system or component(s) using drawings/blueprints and/or system schematics.

    e.   inspection of an aircraft system or component(s) using drawings/blueprints and/or system schematics.

    f.   repair or alteration of an aircraft system or component(s) using drawings/blueprints and/or schematics.

    g.   use of drawings/blueprints in component fabrication.

    h.   terms used in conjunction with aircraft drawings/blueprints and/or system schematics.

2.   N/A

3.   Demonstrates the ability to perform at least one of the following—

    a.   maintenance and/or inspection using drawings/blueprints and/or system schematics. (Level 3)

    b.   preventive maintenance using drawings/blueprints and/or schematics. (Level 3)

    c.   troubleshooting using drawings/blueprints and/or schematics. (Level 3)

    d.   use a control cable tension chart. (Level 3)

    e.   use a servicing, limitation, or calculation chart or graph. (Level 3)

    f.   draw a sketch of an alteration or repair. (Level 2)

    g.   draw a diagram of an electrical circuit or other system, or portion thereof, and explain the drawing. (Level 2)

C. WEIG-HT AND BALANCE

*Core competency element.

REFERENCES: ABS; AMT-G; FAA-H-8083-1.

Objective.   To determine that the applicant:

1. Exhibits knowledge of at least two of the following—

    a. the purpose(s) of weighing or reweighing.
    b. general preparations for weighing, with emphasis on aircraft preparation and/or weighing area considerations.
    c. the general location of airplane center of gravity (CG) in relation to the center of lift for most fixed main airfoils.
    d. definitions of any of the following: datum, arm, moment (positive or negative), or moment index.
    e. the meaning and/or application of any terms/nomenclature associated with weight and balance other than those mentioned in element "d" above, including but not limited to any of the following: tare, ballast, and residual fuel/oil.
    f. procedures for finding any of the following: datum, arm, moment (positive or negative), or moment index.
    g. purpose and/or application of mean aerodynamic chord (MAC).
    h. adverse loading considerations.

2. *Demonstrates the ability to calculate weight and balance CG and complete aircraft weight and balance documentation. (Level 3)

3. Demonstrates the ability to perform at least one of the following—

    a. weighing equipment preparation and setup according to manufacturer's instructions. (Level 3)
    b. locate procedures for leveling and the leveling points for an aircraft. (Level 2)
    c. locate weigh points, procedures for determining CG, and determine the weigh point arms for an aircraft. (Level 2)
    d. identify tare items for a specific aircraft and weighing procedure. (Level 2)
    e. find the datum for at least two different aircraft. (Level 2)
    f. determine the weight and location of required ballast after an (actual or hypothetical) equipment change. (Level 2)

D. FLUID LINES AND FITTINGS

*Core competency element.

REFERENCES: AMT-G; ABS.

Objective.   To determine that the applicant:

| 1.   Exhibits knowledge of at least two of the following— |

      a.   tubing materials.
      b.   tubing materials application.
      c.   tubing sizes.
      d.   flexible hose material.
      e.   flexible hose materials application.
      f.   flexible hose sizes.
      g.   flexible hose identification.
      h.   AN, MS, and/or AC plumbing fittings.
      i.   rigid line fabrication techniques/practices.
      j.   rigid line installation techniques/practices.
      k.   flexible hose fabrication techniques/practices.
      l.   flexible hose installation techniques/practices.

2.   *Demonstrates the ability to perform at least one of the following—

      a.   rigid line fabrication to include tube fittings, bending, and tube flaring. (Level 3)
      b.   flexible line fabrication using replaceable fittings on at least one end. (Level 3)

3.   Demonstrates the ability to perform at least one of the following—

      a.   inspect for and identify defects in rigid and/or flexible lines. (Level 3)
      b.   install and remove a rigid and/or flexible line. (Level 3)
      c.   identify correct and/or incorrect rigid line installations. (Level 2)
      d.   identify correct and/or incorrect flexible line installations. (Level 2)
      e.   form a bead on tubing. (Level 3)
      f.   select components and assemble a flareless fitting tube connection. (Level 3)
      g.   repair a damaged rigid line. (Level 3)
      h.   identify various sizes and types of aircraft fittings. (Level 2)
      i.   secure a rigid line with clamps. (Level 3)
      j.   identify fluid and/or air lines that may be installed on an aircraft. (Level 2)

E. MATERIALS AND PROCESSES

*Core competency element.

REFERENCES: ABS; AMR; AMT-G; JSAT; JSGT.

Objective.   To determine that the applicant:

| 1.   Exhibits knowledge of at least two of the following— |

    a.   any of the metals commonly used in aircraft and their general application. |
    b.   composites and other nonmetallic components and their general application.
    c.   heat-treated parts precautions, using DD or "icebox" rivets.
    d.   typical wood materials and fabric coverings.
    e.   visible characteristics of acceptable and/or unacceptable welds. |
    f.   precision measurement and precision measurement tools.
    g.   using inspection techniques/methods, including any of the following: visual, metallic ring test, dye/fluorescent penetrant, magnetic particle, and/or eddy current.
    h.   identification, selection, installation, and/or use of aircraft hardware.
    i.   safetying of components and/or hardware.
    j.   finding information about material types for specific application(s).

2.   *Demonstrates the ability to torque to specification(s), and safety-wire aircraft component(s)/hardware. (Level 3)

3.   Demonstrates the ability to perform at least one of the following—

    a.   select and install standard aircraft hardware, to include one or more self-locking nuts. (Level 3)
    b.   select, install, and secure a clevis bolt and associated hardware. (Level 3)
    c.   select and install one or more appropriate screws/bolts, nuts, cotter pins, and washers. (Level 3)
    d.   inspect hardware for defects, proper installation. (Level 3)
    e.   safety a turnbuckle. (Level 3)
    f.   perform a dye or fluorescent penetrant inspection. (Level 3)
    g.   find a (not visible) defect using eddy current or ultrasonic inspection equipment. (Level 2)

    h. perform, read, and record a precision measurement using a dial indicator, or micrometer, or vernier caliper. (Level 2)

    i. visually inspect welds and determine acceptability. (Level 3)

    j. identify rivets by physical characteristics. (Level 2)

F. GROUND OPERATION AND SERVICING

REFERENCES: ABS; AMT-G; JSGT.

Objective.   To determine that the applicant:

1. Exhibits knowledge of at least two of the following—

    a. general procedures for towing aircraft.
    b. Air Traffic Control (ATC) considerations/requirements for towing aircraft on or across active runways.
    c. general procedures for starting, ground operating, and/or taxiing a reciprocating engine powered aircraft.
    d. general procedures for starting, ground operating, and/or taxiing a turbine engine powered aircraft.
    e. the hazards associated with starting, ground operating, and/or taxiing aircraft and procedures for preventing, minimizing or otherwise managing any of them.
    f. procedures for refueling and/or defueling aircraft.
    g. oxygen system safety practices/precautions.
    h. characteristics of aviation gasoline and/or turbine fuels, including basic types and means of identification.
    i. fuel contamination hazards.
    j. fuel additives commonly used in the field.
    k. use of automobile fuel in aircraft engines.
    l. types/classes of fires, using proper fire extinguishers/methods.

2. N/A

3. Demonstrates the ability to perform at least one of the following—

    a. service an aircraft with compressed air or nitrogen. (Level 3)
    b. set-up an aircraft and cockpit controls for engine start. (Level 2)
    c. start and ground operate an aircraft engine* (taxiing optional), and use or respond to standard hand or light wand signals. (Level 3)
    d. determine the engine oil for a specific engine. (Level 2)
    e. secure an aircraft for outside storage. (Level 3)

    f.   fuel and/or defuel an aircraft (may be simulated). (Level 3)

    g.   sample fuel and inspect for proper fuel and contaminates. (Level 3)

    h.   set-up and connect an aircraft to an external power source. (Level 2)

    i.   connect a towbar to an aircraft and prepare for towing. (Level 3)

    j.   direct the movement (may be simulated) of aircraft. (Level 3)

    k.   locate and clear a liquid lock (actual or simulated) in an aircraft engine. (Level 3)

    l.   identify the types/classes of fires that local shop and/or flightline fire extinguishers may be used on. (Level 2)

*If an operable engine is available.

G. CLEANING AND CORROSION CONTROL

*Core competency element.

REFERENCES: ABS; AC 43-4A; AMT-G.

Objective.   To determine that the applicant:

  1.   Exhibits knowledge of at least two of the following—

      a.   aircraft preparation for washing, general aircraft cleaning (washing) procedures.

      b.   postcleaning (washing) procedures.

      c.   corrosion theory.

      d.   types/effects of corrosion.

      e.   conditions that cause corrosion.

      f.   corrosion prone areas in aircraft.

      g.   corrosion preventive maintenance procedures.

      h.   inspection for and identification of corrosion in any of its various forms.

      i.   corrosion removal and treatment procedures.

      j.   use of Material Safety Data Sheets (MSDS).

  2.   *Demonstrates the ability to inspect for and identify two or more of the various forms of corrosion that affect aircraft. (Level 3)

  3.   Demonstrates the ability to perform at least one of the following:

      a.   identify and select materials used to clean interior and/or exterior surfaces according to aircraft manufacturer's instructions. (Level 2)

      b.   corrosion removal from any of the metals commonly used in aircraft. (Level 3)

    c.   preventive corrosion treatment on any of the metals commonly used in aircraft. (Level 3)

    d.   identify and select appropriate corrosion preventive methods and materials for a specific aircraft application. (Level 2)

H. MATHEMATICS

REFERENCES: AC 65-9A; ABS; AMT-G.

Objective.   To determine that the applicant:

1.   Exhibits knowledge of at least two of the following—

    a.   areas of various geometrical shapes.
    b.   volumes of various geometrical shapes.
    c.   definitions/descriptions of geometrical terms, including but not limited to any of the following: polygon, pi, diameter, radius, and hypotenuse.
    d.   ratio problems, including one or more examples of where or how they may be used in relation to aircraft maintenance or system(s) operation.
    e.   proportion problems, including one or more examples of where or how they may be used in relation to aircraft maintenance or system(s) operation.
    f.   percentage problems, including one or more examples of where or how they may be used in relation to aircraft maintenance or system(s) operation.
    g.   algebraic operations, including one or more examples of where or how they may be used in relation to aircraft maintenance.
    h.   conditions or areas where metric conversion may be necessary.

2.   N/A

3.   Demonstrates the ability to perform at least one of the following, using appropriate formulas—

    a.   calculate the area of a polygon and/or circle. (Level 2)
    b.   calculate the volume of a sphere, cube, or cylinder. (Level 2)
    c.   algebraic operations involving addition, subtraction, multiplication, and/or division of positive and negative numbers. (Level 2)
    d.   locate mathematical formulas used to assist in the maintenance, preventive maintenance, or alteration of aircraft. (Level 1)

NOTE: The practical portion of the Mathematics subject area may be tested simultaneously when performing calculation(s) in subject areas Basic Electricity and/or Weight and Balance.

I. MAINTENANCE FORMS AND RECORDS

*Core competency element.

REFERENCES: 14 CFR parts 1, 43, and 91.

Objective. To determine that the applicant:

1. Exhibits knowledge of at least two of the following—

   a. writing descriptions of work performed and approval for return to service after minor repairs or minor alterations.
   b. the content, form, and disposition of aircraft maintenance records reflecting approval for return to service after a 100-hour inspection.
   c. the content, form, and disposition of aircraft maintenance records reflecting disapproval for return to service after a 100-hour inspection.
   d. the recording content, form, and disposition requirements for certificated aviation mechanics (without an Inspection Authorization) who perform major repairs and/or major alterations.
   e. the inoperative instruments or equipment provisions of 14 CFR part 91.
   f. the definition/explanation of any of the terms used in relation to aircraft maintenance, such as overhaul(ed), rebuilt, time in service, maintenance, preventive maintenance, inspection, major alteration, major repair, minor alteration, and minor repair.

2. *Demonstrates the ability to write appropriate entries on FAA Form 337, Major Repair and Major Alteration, indicating performance of a major repair, and make appropriate corresponding aircraft maintenance record entry. (Level 3)

3. Demonstrates the ability to write entries for at least one of the following—

   a. performance of minor repair or minor alteration. (Level 3)
   b. performance of preventive maintenance. (Level 3)
   c. compliance with an airworthiness directive. (Level 3)
   d. performance of a 100-hour inspection with approval for return to service, including a list of some allowable inoperative instruments or equipment in accordance with the provision of 14 CFR part 91. (Level 3)

Change 2 (9/24/03)

e. performance of a 100-hour inspection with disapproval for return to service because of needed maintenance, or noncompliance with applicable specifications or airworthiness directive(s). (Level 3)

f. FAA Form 337, Major Repair and Major Alteration, for additional equipment installation or an alteration in accordance with a supplemental type certificate (STC) and make appropriate maintenance record entry. (Level 3)

g. FAA Form 8010-4, Malfunction or Defect Report. (Level 3)

J. BASIC PHYSICS

REFERENCES: ABS; AC 65-15A.

Objective. To determine that the applicant:

1. Exhibits knowledge of at least two of the following—

a. any of the simple machines, how they function, and/or how mechanical advantage is applied in one or more specific examples.

b. sound resonance, how it can be a hazard to aircraft, and how sound may be used to aid in inspecting aircraft.

c. the relationship between fluid density and specific gravity.

d. the characteristic of specific gravity of fluids and how it may be applied to aircraft maintenance.

e. the general effects of pressure and temperature on gases and liquids and how the qualities of compressibility and/or incompressibility of gases and liquids are generally applied to aircraft systems.

f. density altitude and the effects of temperature, and/or pressure, and/or humidity on aircraft and/or engine performance.

g. heat, how it is manifested in matter, and how heat transfer is accomplished through conduction, and/or convection, and/or radiation.

h. coefficient of linear (thermal) expansion as related to aircraft materials.

i. aircraft structures and theory of flight/physics of lift.

j. the operation of aerodynamic factors in the flight of airplanes and/or helicopters.

k. the relationship between force, area, and pressure.

l. the five forces or stresses affecting aircraft structures.

m. the two forms of energy and how they apply to aircraft and/or aircraft systems.

2. N/A

3. Demonstrates the ability to perform at least one of the following—

   a. identify any parts or systems of an aircraft and/or engine where Bernoulli's principle and/or Newtonian law is applied. (Level 2)

   b. identify parts or systems of an aircraft where Boyle's, Charles', and/or Pascal's Laws apply. (Level 2)

   c. calculate force, area, or pressure in a specific application. (Level 3)

   d. identify one or more methods of heat transfer in aircraft systems and where and how heat damage may occur when performing aircraft maintenance. (Level 2)

   e. identify any of the following and describe how they function aerodynamically: stall strips, wing fences, vortex generators, flaps, slats, spoilers, ailerons, stabilators, elevators, rudders, or trim tabs. (Level 2)

   f. determine which of the five forces/stresses are acting on an aircraft or aircraft parts at specific points under given conditions. (Level 2)

   g. design a simple machine (on paper) that uses one or more methods of mechanical advantage. (Level 2)

K. MAINTENANCE PUBLICATIONS

*Core competency element.

REFERENCE: ABS.

Objective. To determine that the applicant:

1. Exhibits knowledge of at least two of the following—

   a. how a mechanic makes use of Type Certificate Data Sheets (TCDSs) and/or Aircraft Specifications in conducting maintenance or inspections.

   b. aircraft maintenance manuals and associated publications including any of the following types of publications and how they are used: service bulletin, maintenance manual, overhaul manual, structural repair manual, or instructions for continued airworthiness.

   c. the requirements of 14 CFR parts 43.13, 43.15, or 43.16 in the performance of maintenance.

   d. Airworthiness Directives (AD), including purpose and/or AD categories and/or ADs issued to other than aircraft.

   e. in what form individuals may receive FAA published AD summaries and/or how they may be obtained.

   f. the AD identification numbering system.

g. FAA Advisory Circulars (ACs) including any of the following: significance of the AC numbering system, one or more examples of ACs issued to provide information in designated subject areas, one or more examples of ACs issued to show a method acceptable to the FAA complying with the CFRs.
h. the intent or function of the Aviation Maintenance Alerts.
i. the Air Transport Association (ATA) Specification 100.

2. *Demonstrates the ability to perform both of the following—

a. read, comprehend, and apply information contained in a manufacturer's maintenance manual or illustrated parts manual. (Level 3)
b. locate and list all applicable ADs for at least one particular make, model, and serial number of an aircraft, engine, propeller, or appliance. (Level 2)

3. Demonstrates the ability to read, comprehend, and apply the information contained in at least one of the following—

a. service bulletin. (Level 3)
b. overhaul manual. (Level 3)
c. structural repair manual. (Level 3)
d. instructions for continued airworthiness. (Level 3)
e. at least one maintenance related section, or appendix, or portion(s) thereof, of 14 CFR. (Level 3)
f. an AD. (Level 3)
g. Aircraft Specifications or TCDSs to specific maintenance or inspection operations, or portions thereof. (Level 3)

L. AVIATION MECHANIC PRIVILEGES AND LIMITATIONS

REFERENCES: 14 CFR part 65; AC 65-30A.

Objective: To determine that the applicant:

1. Exhibits knowledge of mechanic privileges and limitations and exercise thereof, including at least two of the following—

a. required evidence of eligibility experience satisfactory to the Administrator.
b. length of experience required for eligibility.
c. practical experience required for eligibility.
d. the privileges of a mechanic in relation to 100-hour and annual inspections.
e. change of address reporting requirements.
f. minimum age requirements.

g. recent experience requirements to exercise privileges of a certificate.
h. who is authorized to perform maintenance/inspection, preventive maintenance, rebuilding, or alteration and/or approve for return to service afterwards.
i. causes for revocation or suspension.
j. criteria for determining major and minor repair or alteration.

2. N/A

3. When given a copy of 14 CFR part 65, demonstrates the ability to understand mechanic privileges and limitations by finding and interpreting/explaining essential information contained in at least two of the following—

a. Offenses involving alcohol or drugs. (Level 2)
b. Written tests: Cheating or other unauthorized conduct. (Level 2)
c. Applications, certificates, logbooks, reports, and records: falsification, reproduction, or alteration. (Level 2)
d. Refusal to submit to a drug or alcohol test. (Level 2)
e. General privileges and limitations. (Level 2)
f. Recent experience requirements. (Level 2)
g. Airframe rating; additional privileges and/or Powerplant rating; additional privileges. (Level 2)
h. Display of certificate. (Level 2)

**U.S. Department
of Transportation**

**Federal Aviation
Administration**

FAA-S-8081-27
w/ Changes 1, 2, & 3

AVIATION MECHANIC
AIRFRAME

Practical Test Standards

June 2003

Supplied by:

JEPPESEN SANDERSON, INC.
55 Inverness Drive East
Englewood, CO 80112-5498

FLIGHT STANDARDS SERVICE
Washington, DC 20591

NOTE

FAA-S-8081-27, Aviation Mechanic Airframe Practical Test Standards (PTS) is to replace the oral and practical test guides currently used. Both testing procedures will be in effect until all examiners have been trained to administer the test in accordance with the PTS, or 2 years after the effective date of Order 8610.4J, Aviation Mechanic Examiner Handbook. After which time, **all** tests must be administered under the PTS guidelines. New examiners must use the PTS upon completion of initial training. Previously appointed examiners must transition to the PTS within 60 days after completion of recurrent training.

Record of Changes

Change 1: 8/8/2003

Introduction
Performance Levels

> LEVEL 1—Z3b. Nondestructive changed to specified.
> LEVEL 1—PERFORMANCE STANDARD deleted.
> LEVEL 2—bullet 2: added additional text.
> LEVEL 2—PERFORMANCE STANDARD deleted.
> LEVEL 3—bullet 4: added additional text.
> LEVEL 3—Z3e. Verify changed to check.
> LEVEL 3—PERFORMANCE STANDARD deleted.

Change 2: 9/24/2003

Introduction
Performance Levels

> LEVEL 1—PERFORMANCE STANDARD added.
> LEVEL 2—PERFORMANCE STANDARD added.
> LEVEL 3—PERFORMANCE STANDARD added.

Section II—Airframe Structures
A. Wood Structures, Objective 1. Change "at least four" to "at least two."
B. Aircraft Covering, Objective 1. Change "at least four" to "at least two."
C. Aircraft Finishes, Objective 1. Change "at least four" to "at least two."
D. Sheet Metal and Non-metallic Structures, Objective 1. Change "at least four" to "at least two."
E. Welding, Objective 1. Change "at least four" to "at least two."
F. Assembly and Rigging, Objective 1. Change "at least four" to "at least two."
G. Airframe Inspection, Objective 1. Change "at least four" to "at least two."

Section III—Airframe Systems and Components

K. Aircraft Landing Gear Systems, Objective 1. Change "at least four" to "at least two."
L. Hydraulic and Pneumatic Power Systems, Objective 1. Change "at least four" to "at least two."
M. Cabin Atmosphere Control Systems, Objective 1. Change "at least four" to "at least two."
N. Aircraft Instrument Systems, Objective 1. Change "at least four" to "at least two."
O. Communication and Navigation systems, Objective 1. Change "at least four" to "at least two."

P. Aircraft Fuel Systems, Objective 1. Change "at least four" to "at least two."

Q. Aircraft Electrical Systems, Objective 1. Change "at least four" to "at least two."

R. Position and Warning System, Objective 1. Change "at least four" to "at least two."

S. Ice and Rain Control Systems, Objective 1. Change "at least four" to "at least two."

T. Fire Protection Systems, Objective 1. Change "at least four" to "at least two."

Change 3: 6/21/2004

Introduction

Unsatisfactory Performance

Section III—Airframe Systems and Components

G. Airframe Inspection

K. Aircraft Landing Gear Systems

L. Hydraulic and Pneumatic Power Systems

M. Cabin Atmosphere Control Systems

N. Aircraft Instrument Systems

FOREWORD

This Aviation Mechanic Airframe Practical Test Standards book has been published by the Federal Aviation Administration (FAA) to establish the standards for the Aviation Mechanic Airframe Practical Test. The passing of this practical test is a required step toward obtaining the Aviation Mechanic certificate with an Airframe rating. **FAA inspectors and Designated Mechanic Examiners (DMEs) shall conduct practical tests in compliance with these standards.** Applicants should find these standards helpful in practical test preparation.

/s/ 2-13-2003
_____

Joseph K. Tintera, Manager
Regulatory Support Division
Flight Standards Service

CONTENTS

Introduction.. 1

    Practical Test Standard Concept 2
    Practical Test Book Description 2
    Practical Test Standard Description 2
    Use of the Practical Test Standards................................. 4
    Aviation Mechanic Practical Test Prerequisites 4
    Examiner Responsibility... 4
    Performance Levels 5
    Satisfactory Performance.. 6
    Unsatisfactory Performance.. 6

SUBJECT AREAS

SECTION II—AIRFRAME STRUCTURES

    A. WOOD STRUCTURES...2-1
    B. AIRCRAFT COVERING 2-2
    C. AIRCRAFT FINISHES 2-3
    D. SHEET METAL AND NON-METALLIC STRUCTURES 2-4
    E. WELDING 2-5
    F. ASSEMBLY AND RIGGING 2-6
    G. AIRFRAME INSPECTION 2-7
    H. RESERVED 2-8
    I. RESERVED 2-8
    J. RESERVED 2-8

SECTION III—AIRFRAME SYSTEMS AND COMPONENTS

    K. AIRCRAFT LANDING GEAR SYSTEMS............................3-1
    L. HYDRAULIC AND PNEUMATIC POWER SYSTEMS........3-2
    M. CABIN ATMOSPHERE CONTROL SYSTEMS 3-3
    N. AIRCRAFT INSTRUMENT SYSTEMS 3-4
    O. COMMUNICATION AND NAVIGATION SYSTEMS...........3-5
    P. AIRCRAFT FUEL SYSTEMS..3-6
    Q. AIRCRAFT ELECTRICAL SYSTEMS..............................3-7
    R. POSITION AND WARNING SYSTEM 3-8
    S. ICE AND RAIN CONTROL SYSTEMS 3-9
    T. FIRE PROTECTION SYSTEMS 3-10

INTRODUCTION

The Flight Standards Service of the Federal Aviation Administration (FAA) has developed this practical test book as a standard to be used by FAA inspectors and Designated Mechanic Examiners (DMEs) when conducting aviation mechanic practical tests. Applicants are expected to use this book when preparing for practical testing.

Information considered directive in nature is described in this practical test book in terms, such as "shall" and "must" indicating the actions are mandatory. Guidance information is described in terms, such as "should" and "may" indicating the actions are desirable or permissive but not mandatory.

The FAA gratefully acknowledges the valuable assistance provided by the many individuals and organizations who contributed their time and talent in assisting with the development of these practical test standards.

This practical test standard may be downloaded from the Regulatory Support Division's, AFS-600, web site at http://afs600.faa.gov. Subsequent changes to this standard, in accordance with AC 60-27, Announcement of Availability: Changes to Practical Test Standards, will also be available on AFS-600's web site and then later incorporated into a printed revision.

This publication can be purchased from the Superintendent of Documents, U.S. Government Printing Office, Washington, DC 20402. The official online bookstore web site for the U.S. Government Printing Office is www.access.gpo.gov

Comments regarding this document should be sent to:

U.S. Department of Transportation
Federal Aviation Administration
Regulatory Support Division
Airman Testing Standards Branch, AFS-630
P.O. Box 25082
Oklahoma City, OK 73125

Practical Test Standard Concept

Title 14 of the Code of Federal Regulations (14 CFR) specifies the subject areas in which knowledge and skill must be demonstrated by the applicant before the issuance of an Aviation Mechanic Certificate with an Airframe rating. The CFRs provide the flexibility that permits the FAA to publish practical test standards containing knowledge and skill specifics in which competency must be demonstrated.

"Knowledge" (oral) elements are indicated by use of the words *"Exhibits knowledge of...."*

"Skill" (practical) elements are indicated by the use of the words *"Demonstrates the ability to...."*

The FAA will revise these standards whenever it is determined that changes are needed. **Adherence to the applicable regulations, the policies set forth in the current revision of FAA Order 8610.4, Aviation Mechanic Examiner Handbook, and the practical test standards, is mandatory for the evaluation of aviation mechanic applicants.**

Practical Test Book Description

This test book contains the following Aviation Mechanic Airframe Practical Test Standards.

Section II   Airframe Structures
Section III   Airframe Systems and Components

Practical Test Standard Description

The Aviation Mechanic Airframe Practical Test Standards include the subject areas of knowledge and skill for the issuance of an aviation mechanic certificate and/or the addition of a rating. The subject areas are the topics in which aviation mechanic applicants must have knowledge and/or demonstrate skill.

The REFERENCE identifies the publication(s) that describe(s) the subject area. Descriptions of the subject area are not included in the practical test standards, because this information can be found in references listed and/or in manufacturer or FAA-approved or acceptable data related to each subject area. Publications other than those listed may be used as references if their content conveys substantially the same information as the referenced publications. Except where appropriate, (e.g., pertinent CFRs) references listed in this document are NOT meant to supersede or otherwise replace manufacturer or other FAA-approved or acceptable data, but to serve as general information and study material sources.

Information contained in manufacturer and/ or FAA-approved/acceptable data always takes precedence over advisory or textbook referenced data. Written instructions given to applicants for the completion of assigned skill portions of the practical test standard may include service bulletins, airworthiness directives or other CFRs; type certificate data sheets or specifications; manufacturer maintenance manuals or other similar approved/acceptable data necessary for accomplishment of objective testing.

Reference List

14 CFR part 21	Certification Procedures for Products and Parts
14 CFR part 43	Maintenance, Preventive Maintenance Rebuilding, and Alteration
14 CFR part 45	Identification and Registration Markings
14 CFR part 65	Certification: Airmen Other Than Flight Crewmembers
14 CFR part 91	Air Traffic and General Operating Rules
JSAT	A & P Technician Airframe Textbook, Jeppesen Sandersen, Inc.
JSGT	A & P Technician General Textbook, Jeppesen Sandersen, Inc.
ABS	Aircraft Basic Science, Glencoe—Macmillan/McGraw-Hill Publishing Co.
AC 43.13-1	Acceptable Methods, Techniques and Practices—Aircraft Inspection and Repair
AC 65-9	Airframe and Powerplant Mechanics General Handbook
AC 65-15	Airframe and Powerplant Mechanics Airframe Handbook
AMT-A	Aviation Maintenance Technician Series–Airframe, Aviation Supplies and Academics (ASA), Inc.
AMT-G	Aviation Maintenance Technician Series—General, Aviation Supplies and Academics (ASA), Inc.
AMR	Aircraft Maintenance and Repair, Glencoe—Macmillian/MacGraw-Hill Publishing Co.
AEE	Aircraft Electricity and Electronics, Glencoe—Macmillan/McGraw-Hill Publishing Co.

Each subject area has an objective. The objective lists the important knowledge and skill elements that must be utilized by the examiner in planning and administering aviation mechanic tests, and that applicants must be prepared to satisfactorily perform.

EXAMINER is used in this standard to denote either the FAA Inspector or FAA Designated Mechanic Examiner (DME) who conducts the practical test.

Use of the Practical Test Standards

The FAA requires that all practical tests be conducted in accordance with the appropriate Aviation Mechanic Practical Test Standards and the policies and procedures set forth in the current revision of FAA Order 8610.4. When using the practical test book, the examiner must evaluate the applicant's knowledge and skill in sufficient depth to determine that the objective for each subject area element selected is met.

An applicant is not permitted to know before testing begins which selections in each subject area are to be included in his/her test (except the core competency elements, which all applicants are required to perform). Therefore, an applicant should be well prepared in *all* oral and skill areas included in the practical test standard.

Further information about the requirements for conducting/taking the practical test is contained in FAA Order 8610.4

Aviation Mechanic Practical Test Prerequisites

All applicants must have met the prescribed experience requirements as stated in 14 CFR part 65, section 65.77. (See FAA Order 8610.4 for information about testing under the provisions of 14 CFR part 65, section 65.80.)

Examiner Responsibility

The examiner who conducts the practical test is responsible for determining that the applicant meets acceptable standards of knowledge and skill in the assigned subject areas within the appropriate practical test standard. Since there is no formal division between the knowledge and skill portions of the practical test, this becomes an ongoing process throughout the test.

The following terms may be reviewed with the applicant prior to, or during, element assignment.

1. "Inspect" means to examine by sight and/or touch (with or without inspection enhancing tools/equipment).
2. "Check" means to verify proper operation.
3. "Troubleshoot" means to analyze and identify malfunctions.
4. "Service" means to perform functions that assure continued operation.
5. "Repair" means to correct a defective condition.

Performance Levels

The following is a detailed description of the meaning of each level.

Level 1

- Know basic facts and principles.
- Be able to find information and follow directions and written instructions.
- Locate methods, procedures, instructions, and reference material.
- Interpretation of information not required.
- No skill demonstration is required.

Example:

Z3b. Locate specified nondestructive testing methods. (Level 1)

Performance Standard: The applicant will locate information for nondestructive testing.

Level 2

- Know and understand principles, theories, and concepts.
- Be able to find and interpret maintenance data and information, and perform basic operations using appropriate data, tools, and equipment.
- A high level of skill is not required.

Example:

Z3c. Detect electrical leakage in electrical connections, terminal strips, and cable harness (at least 10 will have leakage faults). (Level 2)

Performance Standard: Using appropriate maintenance data and a multimeter, the applicant will identify items with leakage faults.

Level 3

- Know, understand, and apply facts, principles, theories, and concepts.
- Understand how they relate to the total operation and maintenance of aircraft.
- Be able to make independent and accurate airworthiness judgments.
- Perform all skill operations to a return-to-service standard using appropriate data, tools, and equipment. Inspections are performed in accordance with acceptable or approved data.
- A fairly high skill level is required.

Example:

Z3e.   Check control surface travel.   (Level 3)

Performance Standard: Using type certificate data sheets and the manufacturer's service manual, the applicant will measure the control surface travel, compare the travel to the maintenance data, and determine if the travel is within limits.

Satisfactory Performance

The practical test is passed if the applicant demonstrates the prescribed proficiency in the assigned elements (core competency and other selected elements) in each subject area to the required standard. Applicants shall not be expected to memorize all mathematical formulas that may be required in the performance of various elements in this practical test standard. However, where relevant, applicants must be able to locate and apply necessary formulas to obtain correct solutions.

Unsatisfactory Performance

If the applicant does not meet the standards of any of the elements performed (knowledge, core competency, or other skill elements), the associated subject area is failed, and thus the practical test is failed. The examiner or the applicant may discontinue testing any time after the failure of a subject area. In any case, the applicant is entitled to credit for only those subject areas satisfactorily completed. See the current revision of FAA Order 8610.4 for further information about retesting and allowable credit for subject areas satisfactorily completed.

Typical areas of unsatisfactory performance and grounds for disqualification include the following.

1. Any action or lack of action by the applicant that requires corrective intervention by the examiner for reasons of safety.
2. Failure to follow acceptable or approved maintenance procedures while performing skill (practical) projects.
3. Exceeding tolerances stated in the maintenance instructions.
4. Failure to recognize improper procedures.
5. The inability to perform to a return to service standard, where applicable.
6. Inadequate knowledge in any of the subject areas.

SECTION II—AIRFRAME STRUCTURES

A. WOOD STRUCTURES

REFERENCES: AC 43.13-1B, AC 65-15A; AMT-A;JSAT.

Objective.   To determine that the applicant:

1. Exhibits knowledge of at least two of the following——

   a. inspection tools for wood structures.
   b. inspection techniques and practices for wood structures.
   c. effects of moisture/humidity on wood.
   d. types and/or general characteristics of wood used in aircraft structures.
   e. permissible substitutes and/or other materials used in the construction and repair of wood structures.
   f. acceptable wood defects.
   g. non-acceptable wood defects.
   h. wood repair techniques and practices.

2. N/A

3. Demonstrates the ability to perform at least one of the following—

   a. inspect aircraft wood structure or wood sample. (Level 3)
   b. inspect a wood repair for airworthiness. (Level 3)
   c. identify and select aircraft quality/acceptable wood. (Level 2)
   d. determine acceptable repairs or limits for one or more specific defects. (Level 2)
   e. locate data for allowable substitute wood material. (Level 1)
   f. determine the allowable species of wood that can be used as a substitute for spruce, and what, if any, dimensional changes are necessary. (Level 2)
   g. locate wood spar and/or rib structure repair procedures. (Level 1)

B. AIRCRAFT COVERING

REFERENCES: AC 65-15A, AC 43-13 1B; AMT-A; JSAT.

Objective.   To determine that the applicant:

| 1.   Exhibits knowledge of at least two of the following— |

     a.   factors used in determining the proper type covering
        material.
     b.   types of approved aircraft covering material.
     c.   seams commonly used.
     d.   covering textile terms.
     e.   structure surface preparation.
     f.   covering methods commonly used.
     g.   covering means of attachment.
     h.   areas on aircraft covering most susceptible to deterioration.
     i.   aircraft covering preservation/restoration.
     j.   inspection of aircraft covering.
     k.   covering repair techniques and practices.

2.   N/A

3.   Demonstrates the ability to perform at least one of the
    following—

     a.   inspect the repair of a damaged covering for airworthiness.
        (Level 3)
     b.   test a finished covering sample to determine acceptability of
        strength. (Level 3)
     c.   determine the minimum fabric strength covering
        requirements for a specific aircraft. (Level 2)
     d.   determine if a covering sample has appropriate identification
        markings. (Level 2)
     e.   determine acceptable repairs for a specific defect. (Level 2)
     f.   determine the classification (major or minor) of a specific
        repair to a fabric-covered surface. (Level 2)
     g.   locate the requirements for repair of a specific fabric
        covering defect. (Level 1)

C. AIRCRAFT FINISHES

REFERENCES: AC 65-15A; AMT; JSAT; 14 CFR part 45.

Objective. To determine that the applicant:

| 1. Exhibits knowledge of at least two of the following— |

    a. protection of airframe structures.
    b. primer materials.
    c. topcoat materials.
    d. surface preparation for a desired finishing material.
    e. effects of ambient conditions on finishing materials.
    f. effects of improper surface preparation on finishing materials.
    g. regulatory requirements for registration markings.
    h. inspection of aircraft finishes.
    i. safety practices/precautions when using finishing materials.
    j. fungicidal, butyrate, and/or nitrate dopes.
    k. finishing materials application techniques and practices.
    l. where necessary, balance considerations after refinishing.

2. N/A

3. Demonstrates the ability to perform at least one of the following—

    a. select appropriate finishing materials for a specific application. (Level 2)
    b. determine preparation necessary for application of finishing materials to a particular surface. (Level 2)
    c. prepare a surface for application of finishing materials. (Level 3)
    d. apply primer and/or topcoat materials. (Level 3)
    e. inspect one or more finished surfaces. (Level 3)
    f. locate appropriate data to use for a specific finishing task. (Level 1)
    g. determine the allowable location and size of registration numbers for a fixed-wing and/or rotorcraft aircraft. (Level 2)

D.   SHEET METAL AND NON-METALLIC STRUCTURES

*Core competency element.

REFERENCES: AC 43-13.1B, AC 65-9A, AC 65-15A; AMT-A; JSAT.

Objective.   To determine that the applicant:

| 1.   Exhibits knowledge of at least two of the following— |

     a.   inspection/testing of sheet metal structures.
     b.   types of sheet metal defects.
     c.   selection of sheet metal.
     d.   layout, and/or forming of sheet metal.
     e.   selection of rivets.
     f.   rivet layout.
     g.   rivet installation.
     h.   inspection/testing of composite structures.
     i.   types of composite structure defects.
     j.   composite structure fiber, core, and/or matrix materials.
     k.   composite materials storage practices and shelf life.
     l.   composite structure repair methods, techniques, and practices.
     m.   window inspection/types of defects.
     n.   window material storage and handling.
     o.   window installation procedures.
     p.   care and maintenance of windows.
     q.   window temporary and/or permanent repairs.
     r.   maintenance safety practices/precautions for sheet metal, and/or composite materials/structures, and/or windows.

2.   *Demonstrates the ability to install and remove at least two each, of two or more types of rivets. (Level 3)

3.   Demonstrates the ability to perform at least one of the following—

     a.   lay out and form sheet metal to given dimensions; include at least one bend. (Level 3)
     b.   determine a rivet lay out pattern. (Level 2)
     c.   visually inspect an unpainted composite surface. (Level 3)
     d.   inspect a composite structure using a non-destructive testing method (in addition to visual). (Level 3)
     e.   select materials and clean a transparent surface. (Level 3)
     f.   inspect a window or windscreen. (Level 3)
     g.   remove one or more minor scratches from a transparent surface. (Level 3)
     h.   determine hole size to use in a sheet metal repair. (Level 2)

i. inspect a sheet metal assembly or repair for airworthiness. (Level 3)
j. drill and countersink and/or dimple sheet metal. (Level 3)
k. identify the fiber-reinforcing materials in at least three laminated composite structure samples. (Level 2)
l. locate data for composite structure damage assessment. (Level 1)

E. WELDING

REFERENCES: AC 43.13-1B, AC 65-9A, AC 65-15A; AMT-A; JSAT.

Objective.   To determine that the applicant:

1. Exhibits knowledge of at least two of the following—

    a. flame welding gasses.
    b. storage/handling of welding gasses.
    c. flame welding practices and techniques.
    d. inert-gas welding practices and techniques.
    e. purpose and types of shielding gasses.
    f. characteristics of acceptable welds.
    g. characteristics of unacceptable welds.
    h. types of steel tubing welding repairs.
    i. procedures for weld repairs.
    j. soldering preparation, types of solder, and/or flux usage.
    k. welding and/or soldering safety practices/precautions.

2. N/A

3. Demonstrates the ability to perform at least one of the following—

    a. ignite a torch, set one or more specified flame patterns, and accomplish proper torch shutdown. (Level 2)
    b. solder a joint or connection. (Level 2)
    c. using aircraft quality materials, weld or braze a joint. (Level 2)
    d. determine the appropriate method/material(s) to use for a specific welding, soldering, or brazing task. (Level 2)
    e. determine the appropriate data to use for a specific welding, soldering, or brazing task. (Level 1)

F.   ASSEMBLY AND RIGGING

*Core competency element.

REFERENCES: AC 65-9A, AC 65-15A; AMT-A; AMT-G; JSAT.

Objective.   To determine that the applicant:

| 1.   Exhibits knowledge of at least two of the following— |

    a.   control cable.
    b.   control cable maintenance.
    c.   cable connectors.
    d.   cable guides.
    e.   control stops.
    f.   push pull tubes.
    g.   torque tubes.
    h.   bell cranks.
    i.   flutter and flight control balance.
    j.   rigging of airplane or rotorcraft flight controls.
    k.   airplane or rotorcraft flight controls and/or stabilizer systems.
    l.   types of rotorcraft rotor systems.
    m.   rotor vibrations.
    n.   rotor blade tracking.
    o.   aircraft jacking procedures.
    p.   jacking safety practices/precautions.

2   *Demonstrates the ability to check and/or set control surface cable tension. (Level 3)

3.   Demonstrates the ability to perform at least one of the following—

    a.   install a control surface. (Level 3)
    b.   check the static balance of a control surface. (Level 3)
    c.   locate the procedures for rigging a helicopter. (Level 1)
    d.   locate helicopter rotor blade tracking procedures. (Level 1)
    e.   identify fixed-wing aircraft rigging adjustment locations. (Level 2)
    f.   locate leveling methods and procedures for a specific aircraft. (Level 1)
    g.   inspect a flight control system for travel and security. (Level 3)
    h.   inspect a primary flight control cable. (Level 3)
    i.   install one or more swaged cable terminals and check with appropriate gage. (Level 3)
    j.   install one or more Nicopress sleeves and check with appropriate gage. (Level 3)

k. check and adjust as necessary a push-pull flight control system. (Level 3)
l. locate jacking points and leveling locations for a specific aircraft. (Level 2)
m. determine the jacking requirements for a particular aircraft. (Level 2)
n. jack an aircraft or portion thereof (e.g., as appropriate for tire/wheel change, or gear retraction). (Level 3)

G. AIRFRAME INSPECTION

*Core competency element.

REFERENCES: AC 65-9A; AMT-A; JSAT.

Objective. To determine that the applicant:

1. Exhibits knowledge of at least two of the following—

    a. one or more required inspections under 14 CFR part 91.
    b. maintenance requirements under 14 CFR part 43.
    c. inspection requirements under 14 CFR part 43.
    d. requirements for complying with airworthiness directives.
    e. compliance with service letters, instructions for continued airworthiness, and/or bulletins.
    f. maintenance record requirements under 14 CFR part 43.
    g. maintenance record requirements under 14 CFR part 91.

2. *Demonstrates the ability to examine an aircraft maintenance record, and determine if inspection and/or maintenance is due. (Level 3)

3. Demonstrates the ability to perform at least one of the following—

    a. accomplish a 14 CFR part 91 required inspection on an airframe portion or component thereof. (Level 3)
    b. inspect an aircraft or portion thereof after maintenance or preventive maintenance. (Level 3)
    c. determine placarding requirements for a specific aircraft and condition. (Level 2)
    d. determine if all required instruments and equipment for specific operating conditions under 14 CFR part 91 are installed in a particular aircraft. (Level 2)
    e. accomplish a conformity inspection on an airframe portion or component thereof and record results. (Level 3)
    f. generate a checklist for conducting a 100-hour airframe inspection on a specific aircraft. (Level 2)

H. **Reserved**

I. **Reserved**

J. **Reserved**

SECTION III—AIRFRAME SYSTEMS AND COMPONENTS

K. AIRCRAFT LANDING GEAR SYSTEMS

*Core competency element.

REFERENCES: AC 65-15A; AMT-A; JSAT.

Objective. To determine that the applicant:

1. Exhibits knowledge of at least two of the following—

    a. landing gear strut servicing/lubrication.
    b. landing gear steering systems.
    c. landing gear retraction/extension systems.
    d. landing gear inspection.
    e. brake assembly inspection.
    f. wheel and tire construction
    g. tire mounting.
    h. wheel and tire inspection.
    i. wheel bearing inspection.
    j. tire storage, care, and/or servicing.
    k. landing gear and/or tire and wheel safety practices/precautions.

2. *Demonstrates the ability to perform inspection of an installed brake for serviceability. (Level 3)

3. Demonstrates the ability to perform at least one of the following—

    a. determine the proper lubricant(s) for a landing gear. (Level 1)
    b. inspect a landing gear or landing gear component(s). (Level 3)
    c. service an oleo strut. (Level 3)
    d. install a brake lining or brake assembly. (Level 3)
    e. clean and inspect wheel bearings. (Level 3)
    f. disassemble, clean as necessary, and inspect a wheel. (Level 3)
    g. select lubricant, and lubricate wheel bearings. (Level 3)
    h. remove and replace/install a wheel and tire assembly on a landing gear. (Level 3)
    i. inspect a wheel and tire assembly, check tire pressure, and service as necessary. (Level 3)
    j. service a nosewheel shimmy damper. (Level 3)
    k. accomplish a landing gear retraction/extension check. (Level 3)

FAA-S-8081-27

l. replace a tire or tube valve core and check for leaks. (Level 3)

L. HYDRAULIC AND PNEUMATIC POWER SYSTEMS

*Core competency element.

REFERENCES: AC 65-15A; AMT-A; JSAT.

Objective. To determine that the applicant:

1. Exhibits knowledge of at least two of the following—

   a. hydraulic and/or pneumatic system, and/or system component(s) function/operation.
   b. servicing, function, and/or operation of accumulators.
   c. types of hydraulic/pneumatic seals and/or fluid/seal compatibility.
   d. hydraulic/pneumatic seal maintenance procedures.
   e. types of hydraulic/pneumatic filters and/or filter operation.
   f. filter maintenance procedures.
   g. pressure regulators and valves.
   h. servicing hydraulic and/or pneumatic systems.
   i. types/identification and/or characteristics of various hydraulics fluids used in aircraft.
   j. hydraulic/pneumatic system safety practices/precautions.

2. *Demonstrates the ability to select and install a hydraulic seal. (Level 3)

3. Demonstrates the ability to perform at least one of the following—

   a. service a pneumatic or hydraulic system filter. (Level 3)
   b. inspect components or portions of a hydraulic or pneumatic system. (Level 3)
   c. locate fluid servicing instructions and identify/select fluid for a particular aircraft. (Level 2)
   d. service a hydraulic reservoir. (Level 3)
   e. troubleshoot a hydraulic or pneumatic system. (Level 3)
   f. repair a hydraulic or pneumatic system defect. (Level 3)
   g. remove and install hydraulic or pneumatic system component(s) and check operation. (Level 3)
   h. service a hydraulic system accumulator. (Level 3)

M. CABIN ATMOSPHERE CONTROL SYSTEMS

REFERENCES: AC 65-15A; AMT-A; JSAT.

Objective.   To determine that the applicant:

1. Exhibits knowledge of at least two of the following—

    a. exhaust heat exchanger and/or system component(s) function, operation, and/or inspection procedures.
    b. combustion heater and/or system component(s) function, operation, and/or inspection procedures.
    c. vapor-cycle system and/or system component(s) operation, servicing and/or inspection procedures.
    d. air-cycle system and/or system component(s) operation and/or inspection procedures.
    e. cabin pressurization and/or system component(s) operation and/or inspection procedures.
    f. types of oxygen systems and/or oxygen system component(s) operation.
    g. oxygen system maintenance procedures.

2. N/A

3. Demonstrates the ability to perform at least one of the following—

    a. inspect and/or troubleshoot an exhaust heat exchanger cabin heat system or system component(s). (Level 3)
    b. inspect and/or troubleshoot a combustion air heater system and/or system component(s). (Level 3)
    c. select proper solution and leak test oxygen system component(s). (Level 3)
    d. inspect and/or troubleshoot an oxygen system and/or system component(s). (Level 3)
    e. check the operation of an oxygen system. (Level 3)
    f. service an oxygen system. (Level 3)
    g. purge an oxygen system. (Level 3)
    h. inspect and/or troubleshoot a vapor cycle cooling system and/or system component(s). (Level 3)
    i. inspect and/or troubleshoot a cabin pressurization system and/or system component(s). (Level 3)
    j. inspect and/or troubleshoot an air cycle machine system and/or system component(s). (Level 3)
    k. locate procedures for protecting a vapor-cycle system from contamination during component replacement. (Level 1)
    l. locate procedures for servicing a vapor-cycle cooling system. (Level 1)
    m. locate procedures for inspecting a cabin outflow valve. (Level 1)

N. AIRCRAFT INSTRUMENT SYSTEMS

REFERENCES: AC 65-15A; AMT-A; JSAT.

Objective.   To determine that the applicant:

| 1. Exhibits knowledge of at least two of the following— |

    a. magnetic compass operation.
    b. magnetic compass swinging procedures.
    c. gyroscopic instrument(s) purpose and operation. |
    d. vacuum/pressure and/or electrically operated instrument |
        system operation. |
    e. vacuum/pressure and/or electricity operated instrument
        system maintenance procedures.
    f. pitot and/or static instruments purpose and operation.
    g. pitot and/or static system operation.
    h. 14 CFR parts 43 and/or 91 requirements for static system
        checks.
    i. aircraft instrument range markings.

2. N/A

3. Demonstrates the ability to perform at least one of the
    following—

    a. remove and install an aircraft instrument. (Level 3)
    b. accomplish a magnetic compass swing. (Level 3)
    c. determine range/limit markings for one or more instruments.
        (Level 2)
    d. remove, inspect, and install one or more vacuum or
        pressure system filters. (Level 3)
    e. determine the proper setting of a vacuum and/or pressure
        system for a particular aircraft. (Level 2)
    f. inspect and/or troubleshoot portions of a vacuum and/or
        pressure and/or electrically operated instrument power
        system. (Level 3)
    g. inspect portions of a pitot-static system. (Level 3)
    h. find barometric pressure using an altimeter. (Level 2)

O. COMMUNICATION AND NAVIGATION SYSTEMS

REFERENCES: AC 65-15; AMT-A; JSAT; 14 CFR part 91.

Objective.   To determine that the applicant:

| 1.   Exhibits knowledge of at least two of the following— |

    a.   14 CFR part 91 emergency locator transmitter (ELT) maintenance requirements.
    b.   14 CFR part 91 ELT record keeping requirements.
    c.   checking/inspecting coaxial cable.
    d.   coaxial cable installation and/or routing requirements.
    e.   communication and/or navigation systems commonly used.
    f.   proper installation of a com/nav radio in an existing radio rack.
    g.   means of identification of commonly used communication and/or navigation antennas.
    h.   autopilot system basic components and/or sensing elements.
    i.   static discharger function and operation.
    j.   static discharger maintenance procedures.

2. N/A

3. Demonstrates the ability to perform at least one of the following—

    a. identify and inspect com/nav cable and connectors. (Level 3)
    b. inspect an ELT and/or ELT installation. (Level 3)
    c. determine ELT battery serviceability/status. (Level 2)
    d. inspect one or more antenna installations. (Level 3)
    e. inspect a coaxial cable installation. (Level 3)
    f. inspect a com/nav radio installation. (Level 3)
    g. inspect a shock mount base. (Level 3)
    h. locate and identify various antennas installed on a particular aircraft. (Level 2)
    i. inspect one or more static dischargers for security, resistance. (Level 3)

P.   AIRCRAFT FUEL SYSTEMS

*Core competency element.

REFERENCES: AC 65-9A; AC 65-15A; AMT-A; JSAT; AMR.

Objective.   To determine that the applicant:

| 1.   Exhibits knowledge of at least two of the following— |

   a.   fuel system strainer servicing.
   b.   construction characteristics of one or more types of fuel
        tanks.
   c.   fuel tank maintenance procedures.
   d.   fuel line routing/installation requirements.
   e.   hazards associated with fuel system maintenance.
   f.   types, characteristics, and/or operation of fuel systems
        and/or components thereof.
   g.   characteristics, and/or operation of fuel jettison systems
        and/or components thereof.

2.   *Demonstrates the ability to service a fuel system strainer.
     (Level 3)

3.   Demonstrates the ability to perform at least one of the
     following—

   a.   install a fuel quantity transmitter and/or accomplish an
        operational check. (Level 3)
   b.   install a fuel valve and/or accomplish an operational check.
        (Level 3)
   c.   install a fuel pump and/or accomplish an operational check.
        (Level 3)
   d.   troubleshoot a fuel system. (Level 3)
   e.   determine the airworthiness of a specified size fuel system
        leak/seep. (Level 2)
   f.   inspect a fuel system and/or fuel system component(s).
        (Level 3)
   g.   check the operation of one or more fuel system
        components. (Level 3)
   h.   inspect a metal fuel tank. (Level 3)
   i.   inspect a bladder fuel tank. (Level 3)
   j.   locate fuel system operating instructions. (Level 1)
   k.   locate fuel system inspection procedures. (Level 1)

Q. AIRCRAFT ELECTRICAL SYSTEMS

*Core competency element.

REFERENCES: AC 65-9A, AC 65-15A; AMT-A; JSAT; JSGT; AEE.

Objective. To determine that the applicant:

| 1. Exhibits knowledge of at least two of the following— |

    a. factors to consider when selecting wire size for an aircraft circuit.
    b. routing and/or installation of electric wire or wire bundles.
    c. wire splicing.
    d. use of derating factors in switch selection.
    e. requirements for circuit protection devices.
    f. voltage regulator—purpose and operating characteristics.
    g. lighting and/or lighting system components.
    h. electric motor operation and/or motor components.
    i. constant speed drive (CSD) and/or integrated drive generator (IDG) systems and/or system components.
    j. airframe electrical system components.
    k. wiring defects and/or inspection.

2. *Demonstrates the ability to troubleshoot an electrical system or portion thereof, using appropriate tools and/or test equipment. (Level 3)

3. Demonstrates the ability to perform at least one of the following–

    a. select a circuit switch or circuit protection device for a specific aircraft and application. (Level 2)
    b. install a circuit switch or circuit protection device. (Level 3)
    c. select materials and tools and accomplish a wire splice. (Level 3)
    d. adjust one or more voltage regulators. (Level 3)
    e. select and install one or more wires and pins and/or sockets in a connector. (Level 3)
    f. select materials and fabricate a bonding wire. (Level 3)
    g. install a bonding wire and accomplish a resistance check. (Level 3)
    h. check the operation of one or more airframe electrical system circuits and/or system components. (Level 3)
    i. inspect and check a landing light. (Level 3)
    j. inspect and check anti-collision and position lights. (Level 3)
    k. inspect generator brushes and determine serviceability. (Level 3)

R. POSITION AND WARNING SYSTEM

REFERENCES: AC 65-15A; AMT-A; JSAT.

Objective. To determine that the applicant:

| 1. Exhibits knowledge of at least two of the following— |

    a. anti-skid system basic components.
    b. anti-skid system operating characteristics.
    c. takeoff warning system basic components.
    d. takeoff warning system function and operation.
    e. control-surface trim indicating system basic components and/or operating characteristics.
    f. landing gear position indicators.
    g. flap position indicators.
    h. landing gear warning system basic components and/or operating characteristics.
    i. checking and/or repairing a landing gear warning system.
    j. types of stall warning/lift detector systems and/or operating characteristics.
    k. common annunciator system indications.
    l. mach warning system indicator(s) and/or operating characteristics.

2. N/A

3. Demonstrates the ability to perform at least one of the following—

    a. inspect and/or adjust a landing gear position switch. (Level 3)
    b. accomplish an operational check of a landing gear position indicating and/or warning system. (Level 3)
    c. inspect and/or adjust a flap position indicating system. (Level 3)
    d. check the operation of a flap position indicating and/or warning system. (Level 3)
    e. troubleshoot a landing gear warning system. (Level 3)
    f. check the operation of an annunciator system. (Level 3)
    g. check the operation of an anti-skid warning system. (Level 3)
    h. identify landing gear position/warning system components. (Level 2)
    i. locate troubleshooting procedures for an anti-skid system. (Level 1)
    j. locate troubleshooting procedures for a landing gear warning system. (Level 1)

S. ICE AND RAIN CONTROL SYSTEMS

REFERENCES: AC 65-15A; JSAT; AMT-A.

Objective.   To determine that the applicant:

1. Exhibits knowledge of at least two of the following—

   a. aircraft icing causes/effects.
   b. ice detection systems.
   c. anti-ice and/or deice areas.
   d. anti-ice and/or deice methods commonly used.
   e. checking and/or troubleshooting a pitot-static anti-ice system.
   f. anti-icing and/or de-icing system components/operation.
   g. anti-icing and/or de-icing system maintenance.
   h. types of rain removal systems and/or operating characteristics.

2. N/A

3. Demonstrates the ability to perform at least one of the following—

   a. troubleshoot a pitot anti-ice system. (Level 3)
   b. check the operation of a pitot-static anti-ice system. (Level 3)
   c. inspect a deicer boot. (Level 3)
   d. check deicer boot operation. (Level 3)
   e. inspect windshield wiper blade(s) and check blade tension. (Level 3)
   f. adjust a windshield wiper blade tension to specification. (Level 3)
   g. inspect an electrically-heated windshield. (Level 3)
   h. check an electrically-heated windshield operation. (Level 3)
   i. troubleshoot a pneumatic deicer boot system. (Level 3)
   j. service or repair on a pneumatic deicer boot. (Level 3)

T. FIRE PROTECTION SYSTEMS

REFERENCES: AC 65-15A; AMT-A; JSAT.

Objective.   To determine that the applicant:

| 1.   Exhibits knowledge of at least two of the following— |

    a. fire and/or smoke detection system(s) or system components.

    b. fire extinguishing system(s) and/or system components.

    c. fire and/or smoke detection system operating characteristics.

    d. fire extinguishing system operating characteristics.

    e. determining proper container pressure for an installed fire extinguisher system.

    f. maintenance procedures for fire detection and/or extinguishing system(s) and/or system component(s).

    g. inspecting and/or checking a fire detection/overheat system.

    h. inspecting and/or checking a smoke and/or toxic gas detection system.

    i. troubleshooting a fire detection and/or extinguishing system.

  2. N/A

  3. Demonstrates the ability to perform at least one of the following—

    a. inspect a fire extinguisher container and determine if the pressure is within limits. (Level 3)

    b. determine the hydrostatic test date of a fire extinguisher container. (Level 2)

    c. troubleshoot a fire detection system. (Level 3)

    d. install/replace one or more smoke and/or fire detection and/or extinguishing system components. (Level 3)

    e. inspect a smoke and/or fire detection and/or extinguishing system, or system component(s). (Level 3)

    f. locate inspection procedures for carbon monoxide detectors. (Level 1)

    g. locate procedures for checking a smoke detection system. (Level 1)

U.S. Department
of Transportation

Federal Aviation
Administration

FAA-S-8081-28
w/ Changes 1 & 2

AVIATION MECHANIC POWERPLANT

Practical Test Standards

June 2003

Supplied by:

JEPPESEN SANDERSON, INC.
55 Inverness Drive East
Englewood, CO 80112-5498

FLIGHT STANDARDS SERVICE
Washington, DC 20591

NOTE

FAA-S-8081-28, Aviation Mechanic Powerplant Practical Test Standards (PTS) is to replace the oral and practical test guides currently used. Both testing procedures will be in effect until all examiners have been trained to administer the test in accordance with the PTS, or 2 years after the effective date of Order 8610.4J, Aviation Mechanic Examiner Handbook. After which time, **all** tests must be administered under the PTS guidelines. New examiners must use the PTS upon completion of initial training. Previously appointed examiners must transition to the PTS within 60 days after completion of recurrent training.

Record of Changes

Change 1: 8/8/2003

Introduction
Performance Levels

LEVEL 1—Z3b. Nondestructive changed to specified.
LEVEL 1—PERFORMANCE STANDARD deleted.
LEVEL 2—bullet 2: added additional text.
LEVEL 2—PERFORMANCE STANDARD deleted.
LEVEL 3—bullet 4: added additional text.
LEVEL 3—Z3e. Verify changed to check.
LEVEL 3—PERFORMANCE STANDARD deleted.

Change 2: 9/24/2003

Introduction
Performance Levels (page 4 & 5)

LEVEL 1—PERFORMANCE STANDARD added.
LEVEL 2—PERFORMANCE STANDARD added.
LEVEL 3—PERFORMANCE STANDARD added.

Section IV—Powerplant Theory and Maintenance

A. Reciprocating Engines, Objective 1. Change "at least four" to "at least two."
B. Turbine Engines, Objective 1. Change "at least four" to "at least two."
C. Engine Inspection, Objective 1. Change "at least four" to "at least two."

Section V—Powerplant Systems and Components

H. Powerplant Systems and Components, Objective 1. Change "at least four" to "at least two."
I. Engine Fire Protection Systems, Objective 1. Change "at least four" to "at least two."
J. Engine Electrical Systems, Objective 1. Change "at least four" to "at least two."
K. Lubrication Systems, Objective 1. Change "at least four" to "at least two."
L. Ignition and Starting Systems, Objective 1. Change "at least four" to "at least two."
M. Fuel Metering Systems, Objective 1. Change "at least four" to "at least two."
N. Engine Fuel Systems, Objective 1. Change "at least four" to "at least two."
O. Induction and Engine Airflow Systems, Objective 1. Change "at least four" to "at least two."

P. Engine Cooling Systems, Objective 1. Change "at least four" to "at least two."

Q. Engine Exhaust and Reverser Systems, Objective 1. Change "at least four" to "at least two."

R. Propellers, Objective 1. Change "at least four" to "at least two.")

S. Turbine Powered Auxiliary Power Units, Objective 1. Change "at least four" to "at least two."

FOREWORD

This Aviation Mechanic Powerplant Practical Test Standards book has been published by the Federal Aviation Administration (FAA) to establish the standards for the Aviation Mechanic Powerplant Practical Test. The passing of this practical test is a required step toward obtaining the Aviation Mechanic certificate with a Powerplant rating. **FAA inspectors and Designated Mechanic Examiners (DMEs) shall conduct practical tests in compliance with these standards.** Applicants should find these standards helpful in practical test preparation.

Joseph K. Tintera, Manager
Regulatory Support Division
Flight Standards Service

CONTENTS

Introduction... 1

    Practical Test Standard Concept 2
    Practical Test Book Description 2
    Practical Test Standard Description.. 2
    Use of the Practical Test Standards... 3
    Aviation Mechanic Practical Test Prerequisites 4
    Examiner Responsibility... 4
    Performance Levels 4
    Satisfactory Performance.. 5
    Unsatisfactory Performance.. 6

SUBJECT AREAS

SECTION IV—POWERPLANT THEORY AND MAINTENANCE

    A.   RECIPROCATING ENGINES 4-1
    B.   TURBINE ENGINES.. 4-2
    C.   ENGINE INSPECTION.. 4-3
    D.   RESERVED... 4-3
    E.   RESERVED... 4-3
    F.   RESERVED... 4-3
    G.   RESERVED... 4-3

SECTION V—POWERPLANT SYSTEMS AND COMPONENTS

    H.   ENGINE INSTRUMENT SYSTEMS 5-1
    I.   ENGINE FIRE PROTECTION SYSTEMS.................... 5-2
    J.   ENGINE ELECTRICAL SYSTEMS 5-3
    K.   LUBRICATION SYSTEMS 5-4
    L.   IGNITION AND STARTING SYSTEMS....................... 5-5
    M.   FUEL METERING SYSTEMS 5-6
    N.   ENGINE FUEL SYSTEMS 5-7
    O   INDUCTION AND ENGINE AIRFLOW SYSTEMS........ 5-8
    P.   ENGINE COOLING SYSTEMS 5-9
    Q.   ENGINE EXHAUST AND REVERSER SYSTEMS 5-10
    R.   PROPELLERS... 5-11
    S.   TURBINE POWERED AUXILIARY POWER UNITS ... 5-12

INTRODUCTION

The Flight Standards Service of the Federal Aviation Administration (FAA) has developed this practical test book document as a standard to be used by FAA inspectors and Designated Mechanic Examiners (DMEs) when conducting aviation mechanic practical tests. Applicants are expected to use this book when preparing for practical testing.

Information considered directive in nature is described in this practical test book in terms, such as "shall" and "must" indicating the actions are mandatory. Guidance information is described in terms, such as "should" and "may" indicating the actions are desirable or permissive, but not mandatory.

The FAA gratefully acknowledges the valuable assistance provided by the many individuals and organizations who contributed their time and experience in assisting with the development of these practical test standards.

This practical test standard may be downloaded from the Regulatory Support Division's, AFS-600, web site at http://afs600.faa.gov. Subsequent changes to this standard, in accordance with AC 60-27, Announcement of Availability: Changes to Practical Test Standards, will also be available on AFS-600's web site and then later incorporated into a printed revision.

This publication can be purchased from the Superintendent of Documents, U.S. Government Printing Office, Washington, DC 20402. The official online bookstore web site for the U.S. Government Printing Office is http://www.access.gpo.gov

Comments regarding this document should be sent to:

U.S. Department of Transportation
Federal Aviation Administration
Regulatory Support Division
Airman Testing Standards Branch, AFS-630
P.O. Box 25082
Oklahoma City, OK 73125

Practical Test Standard Concept

Title 14 of the Code of Federal Regulations (14 CFR) specifies the subject areas in which knowledge and skill must be demonstrated by the applicant before the issuance of an Aviation Mechanic Certificate with a Powerplant rating. The CFRs provide the flexibility that permits the FAA to publish practical test standards containing knowledge and skill specifics in which competency must be demonstrated. The FAA will revise this book whenever it is determined that changes are needed.

> "Knowledge" (oral) elements are indicated by use of the words *"Exhibits knowledge of...."*

> "Skill" (practical) elements are indicated by the use of the words *"Demonstrates the ability to...."*

Adherence to the applicable regulations, the policies set forth in the current revision of FAA Order 8610.4, Aviation Mechanic Examiner Handbook, and the practical test standards, is mandatory for the evaluation of aviation mechanic applicants.

Practical Test Book Description

This test book contains the following Aviation Mechanic Powerplant Practical Test Standards.

> **Section IV—Powerplant Theory and Maintenance**
> **Section V   —Powerplant Systems and Components**

The Aviation Mechanic Powerplant Practical Test Standards include the subject areas of knowledge and skill for the issuance of an aviation mechanic certificate and/or the addition of a rating. The subject areas are the topics in which aviation mechanic applicants must have knowledge and/or demonstrate skill.

Practical Test Standard Description

The REFERENCE identifies the publication(s) that describe(s) the subject area. Descriptions of the subject area are not included in the practical test standards, because this information can be found in references listed and/or in manufacturer or FAA-approved or acceptable data related to each subject area. Publications other than those listed may be used as references if their content conveys substantially the same information as the referenced publications. Except where appropriate, (e.g., pertinent CFRs) references listed in this document are NOT meant to supersede or otherwise replace manufacturer or other FAA-approved or acceptable data, but to serve as general information and study material sources.

Information contained in manufacturer and/ or FAA-approved/acceptable data always takes precedence over advisory or textbook referenced data. Written instructions given to applicants for the completion of assigned skill portions of the practical test standard may include service bulletins; airworthiness directives or other federal aviation regulations; type certificate data sheets or specifications; manufacturer maintenance manuals or other similar approved/ acceptable data necessary for accomplishment of objective testing.

Reference List

14 CFR part 43	Maintenance, Preventive Maintenance Rebuilding and Alteration
AC 43.13-1	Acceptable Methods, Techniques and Practices— Aircraft Inspection and Repair
AC 65-12	Airframe and Powerplant Mechanics Powerplant Handbook
AC 65-15	Airframe and Powerplant Mechanics Airframe Handbook
AGTP	Aircraft Gas Turbine Powerplants, Jeppesen Sandersen, Inc.
AP	Aircraft Powerplants, Glencoe/ McGraw-Hill Publishing Co.
JSPT	A & P Technician Powerplant Textbook, Jeppesen Sandersen, Inc.

Each subject area has an objective. The objective lists the important knowledge and skill elements that must be utilized by the examiner in planning and administering aviation mechanic tests, and that applicants must be prepared to satisfactorily perform.

EXAMINER is used throughout this standard to denote either the FAA Inspector or FAA Designated Mechanic Examiner (DME) who conducts the practical test.

Use of the Practical Test Standards

The FAA requires that all practical tests be conducted in accordance with the appropriate Aviation Mechanic Practical Test Standard, and the policies and procedures set forth in the current revision of FAA Order 8610.4. When using the practical test standards, the examiner must evaluate the applicant's knowledge and skill in sufficient depth to determine that the objective for each subject area element selected is met.

An applicant is not permitted to know before testing begins which selections in each subject area are to be included in his/her test (except the core competency elements, which all applicants are required to perform). Therefore, an applicant should be well prepared in *all* oral and skill areas included in the practical test standard.

Further information about the requirements for conducting/taking the practical test is contained in FAA Order 8610.4

Aviation Mechanic Practical Test Prerequisites

All applicants must have met the prescribed experience requirements as stated in 14 CFR part 65, section 65.77. (See FAA Order 8610.4 for information about testing under the provisions of 14 CFR part 65, section 65.80.)

Examiner Responsibility

The examiner who conducts the practical test is responsible for determining that the applicant meets acceptable standards of knowledge and skill in the assigned subject areas within the appropriate practical test standard. Since there is no formal division between the knowledge and skill portions of the practical test, this becomes an ongoing process throughout the test.

The following terms may be reviewed with the applicant prior to, or during, element assignment.

1. "Inspect" means to examine by sight and/or touch (with or without inspection enhancing tools/equipment).
2. "Check" means to verify proper operation.
3. "Troubleshoot" means to analyze and identify malfunctions.
4. "Service" means to perform functions that assure continued operation.
5. "Repair" means to correct a defective condition.

Performance Levels

The following is a detailed description of the meaning of each level.

Level 1

- Know basic facts and principles.
- Be able to find information and follow directions and written instructions.
- Locate methods, procedures, instructions, and reference material.
- Interpretation of information not required.
- No skill demonstration is required.

Example:

Z3b. Locate specified nondestructive testing methods. (Level 1)

Performance Standard: The applicant will locate information for nondestructive testing.

Level 2

- Know and understand principles, theories, and concepts.
- Be able to find and interpret maintenance data and information, and perform basic operations using appropriate data, tools, and equipment.
- A high level of skill is not required.

Example:

Z3c. Detect electrical leakage in electrical connections, terminal strips, and cable harness (at least 10 will have leakage faults). (Level 2)

Performance Standard: Using appropriate maintenance data and a multimeter, the applicant will identify items with leakage faults.

Level 3

- Know, understand, and apply facts, principles, theories, and concepts.
- Understand how they relate to the total operation and maintenance of aircraft.
- Be able to make independent and accurate airworthiness judgments.
- Perform all skill operations to a return-to-service standard using appropriate data, tools, and equipment. Inspections are performed in accordance with acceptable or approved data.
- A fairly high skill level is required.

Example:

Z3e. Check control surface travel. (Level 3)

Performance Standard: Using type certificate data sheets and the manufacturer's service manual, the applicant will measure the control surface travel, compare the travel to the maintenance data, and determine if the travel is within limits.

Satisfactory Performance

The practical test is passed if the applicant demonstrates the prescribed proficiency in the assigned elements (core competency and other selected elements) in each subject area to the required standard. Applicants shall not be expected to memorize all mathematical formulas that may be required in the performance of various elements in this practical test standard. However, where relevant, applicants must be able to locate and apply necessary formulas to obtain correct solutions.

Unsatisfactory Performance

If the applicant does not meet the standards of any of the elements performed (knowledge, core competency, or other skill elements), the associated subject area is failed, and thus the practical test is failed. The examiner or the applicant may discontinue testing any time after the failure of a subject area. In any case, the applicant is entitled to credit for only those subject areas satisfactorily completed. See the current revision of FAA Order 8610.4 for further information about retesting and allowable credit for subject areas satisfactorily completed.

Typical areas of unsatisfactory performance and grounds for disqualification include the following.

1. Any action or lack of action by the applicant that requires corrective intervention by the examiner for reasons of safety.
2. Failure to follow acceptable or approved maintenance procedures while performing projects.
3. Exceeding tolerances stated in the maintenance instructions.
4. Failure to recognize improper procedures.
5. The inability to perform to a return to service standard, where applicable.
6. Inadequate knowledge in any of the subject areas.

SECTION IV—POWERPLANT THEORY AND MAINTENANCE

A. RECIPROCATING ENGINES

REFERENCES: AC 65-12A; JSPT.

Objective. To determine that the applicant:

| 1.   Exhibits knowledge of at least two of the following— |

    a.   reciprocating engine theory of operation.
    b.   basic radial engine design, components, and/or operation.
    c.   firing order of a reciprocating engine.
    d.   probable cause and removal of a hydraulic lock.
    e.   valve adjustment on a radial engine.
    f.   purpose of master and/or articulating rods.
    g.   checks necessary to verify proper operation of a reciprocating engine.
    h.   induction system leak indications.
    i.   reciprocating engine maintenance procedures.
    j.   procedures for inspecting various engine components during an overhaul.
    k.   correct installation of piston rings and results of incorrectly installed or worn rings.
    l.   purpose/function/operation of various reciprocating engine components, including, but not limited to, any of the following: crankshaft dynamic dampers, multiple springs for valves, piston rings, and reduction gearing.

   2.   N/A

   3.   Demonstrates the ability to perform at least one of the following—

    a.   measure the valve clearance on a reciprocating aircraft engine when the lifters are deflated. (Level 2)
    b.   accomplish a compression test, and note all findings. (Level 3)
    c.   inspect engine control cables and/ or push-pull tubes for proper rigging. (Level 3)
    d.   inspect ring gap, install piston rings on a piston, and install an aircraft engine cylinder. (Level 3)
    e.   dimensionally inspect an aircraft engine component. (Level 3)
    f.   replace/install one or more aircraft engine components. (Level 3)

B. TURBINE ENGINES

REFERENCES: JSPT; AP.

Objective. To determine that the applicant:

| 1. Exhibits knowledge of at least two of the following— |

    a. turbine engine theory of operation.
    b. checks necessary to verify proper operation.
    c. turbine engine troubleshooting procedures.
    d. procedures required after the installation of a turbine engine.
    e. causes for turbine engine performance loss.
    f. purpose/function/operation of various turbine engine components.
    g. turbine engine maintenance procedures.

2. N/A

3. Demonstrates the ability to perform at least one of the following:

    a. repair a turbine engine compressor blade by blending. (Level 3)
    b. remove and/or install a turbine engine component. (Level 3)
    c. determine cycle life remaining between overhaul of a turbine engine life limited component. (Level 2)
    d. check rigging of a turbine engine inlet guide vane system. (Level 3)
    e. measure compressor or turbine blade clearance. (Level 3)
    f. troubleshoot a turbine engine. (Level 3)
    g. locate and identify turbine engine components. (Level 2)
    h. inspect turbine engine components. (Level 3)

NOTE: T. AUXILIARY POWER UNITS may be tested at the same time as AREA B. No further testing of auxiliary power units is required.

C. ENGINE INSPECTION

*Core competency element

REFERENCES: AC 43.13-1B; 14 CFR part 43.

Objective. To determine that the applicant:

1. Exhibits knowledge of at least two of the following—

    a. the use of a type certificate data sheet (TCDS) to identify engine accessories.
    b. requirements for the installation or modification in accordance with a supplemental type certificate (STC).
    c. procedures for accomplishing a 100-hour inspection in accordance with the manufacturer's instruction.
    d. compliance with airworthiness directives.
    e. changes to an inspection program due to a change or modification required by airworthiness directive or service bulletin.
    f. determination of life limited parts.
    g. inspection required after a potentially damaging event, including but not limited to any of the following: sudden stoppage, overspeed, or overtemperature.

2. *Demonstrates the ability to perform inspection of a reciprocating and/or turbine engine installation in accordance with the manufacturer's instructions. (Level 3)

3. Demonstrates the ability to perform at least one of the following—

    a. inspect a turbine engine using a bore scope. (Level 3)
    b. determine proper crankshaft flange run-out. (Level 3)
    c. inspect an engine in accordance with applicable airworthiness directive.
       (Level 2)
    d. inspect a turbine engine compressor section. (Level 3)
    e. inspect a crankcase for cracks. (Level 3)
    f. inspect a crankshaft oil seal for leaks. (Level 3)
    g. engine conformity inspection. (Level 3)
    h. engine airworthiness inspection. (Level 3)

D. Reserved
E. Reserved
F. Reserved
G. Reserved

FAA-S-8081-28

SECTION V—POWERPLANT SYSTEMS AND COMPONENTS

H.   ENGINE INSTRUMENT SYSTEMS

*Core competency element

REFERENCES: AGTP; AC 65-15A.

Objective. To determine that the applicant:

1.   Exhibits knowledge of at least two of the following—

    a.   troubleshoot a fuel flow and/or low fuel pressure indicating system.
    b.   the operation of a fuel flow indicating system and where it is connected to the engine.
    c.   the operation of a temperature indicating system.
    d.   the operation of a pressure indicating system.
    e.   the operation of an RPM indicating system.
    f.   required checks to verify proper operation of a temperature indicating system.
    g.   required checks to verify proper operation of a pressure indicating system.
    h.   required checks to verify proper operation of an RPM indicating system.
    i.   the operation of a manifold pressure gage and where it actually connects to an engine.

2.   *Demonstrates the ability to perform inspection of engine electrical and/or mechanical instrument systems to include at least one of the following (Level 3)—

    a.   temperature.
    b.   pressure.
    c.   RPM.
    d.   rate of flow.

3.   Demonstrates the ability to perform at least one of the following—

    a.   verify proper operation and marking of an indicating system. (Level 2)
    b.   replace a temperature sending unit. (Level 3)
    c.   remove, inspect, and install fuel flow transmitter. (Level 3)
    d.   troubleshoot an oil pressure indicating system. (Level 3)
    e.   locate and inspect fuel flow components on an engine. (Level 2)
    f.

 g. replace an exhaust gas temperature (EGT) indication
  probe. (Level 3)
 h. troubleshoot a manifold pressure gage that is slow to
  indicate the correct reading. (Level 2)

I. ENGINE FIRE PROTECTION SYSTEMS

REFERENCES: AP; JSPT.

Objective. To determine that the applicant:

1. Exhibits knowledge of at least two of the following—

 a. checks to verify proper operation of an engine fire detection
  and/or extinguishing system.
 b. troubleshoots an engine fire detection and/or extinguishing
  system.
 c. inspection requirements for an engine fire extinguisher
  squib and safety practices/precautions.
 d. components and/or operation of an engine fire detection
  and/or extinguishing system.
 e. engine fire detection and/or extinguishing system
  maintenance procedures.

2. N/A

3. Demonstrates the ability to perform at least one of the
 following:

 a. check an engine fire detection and/or extinguishing system
  for proper operation. (Level 2)
 b. accomplish weight and pressure inspection of an engine
  fire bottle, and verify hydrostatic inspection date. (Level 2)
 c. repair an engine fire detector heat sensing loop
  malfunction. (Level 3)
 d. check operation of firewall shut-off valve after a fire handle
  is pulled. (Level 2)
 e. troubleshoot an engine fire detection or extinguishing
  system. (Level 2)
 f. inspect an engine fire detection or extinguishing system.
  (Level 2)

J. ENGINE ELECTRICAL SYSTEMS

REFERENCES: AP; JSPT.

Objective. To determine that the applicant:

| 1.   Exhibits knowledge of at least two of the following— |

      a. generator rating and performance data location.
      b. operation of a turbine engine starter-generator.
      c. the procedure for locating the correct electrical cable/wire size needed to fabricate a replacement cable/wire.
      d. installation practices for wires running close to exhaust stacks or heating ducts.
      e. operation of engine electrical system components.
      f. types of and/or components of D.C. motors.
      g. inspection and/or replacement of starter-generator brushes.

2. N/A

3. Demonstrates the ability to perform at least one of the following—

      a. flash a generator field. (Level 3)
      b. install an engine driven generator or alternator. (Level 3)
      c. use of an engine electrical wiring schematic. (Level 2)
      d. accomplish the installation of a tach generator. (Level 3)
      e. fabricate an electrical system cable. (Level 3)
      f. repair a damaged engine electrical system wire. (Level 3)
      g. replace and check a current limiter. (Level 3)
      h. check/service/adjust one or more engine electrical system components. (Level 3)
      i. troubleshoot an engine electrical system component. (Level 3)

K. LUBRICATION SYSTEMS

REFERENCES: JSPT; AP.

Objective. To determine that the applicant:

1. Exhibits knowledge of at least two of the following—

    a. differences between straight mineral oil, ashless-dispersant oil, and synthetic oil.
    b. types of oil used for different climates.
    c. functions of an engine oil.
    d. identification and selection of proper lubricants.
    e. servicing of the lubrication system.
    f. the reasons for changing engine lubricating oil at specified intervals.
    g. the purpose and operation of an oil/air separator.
    h. reasons for excessive oil consumption without evidence of oil leaks in a reciprocating and/or turbine aircraft engine.

2. N/A

3. Demonstrates the ability to perform at least one of the following—

    a. inspect an engine lubrication system to ensure continued operation. (Level 3)
    b. inspect oil lines and filter/screen for leaks. (Level 3)
    c. replace a defective oil cooler or oil cooler component. (Level 3)
    d. replace a gasket or seal in the oil system, and accomplish a leak check. (Level 3)
    e. adjust oil pressure. (Level 3)
    f. change engine oil, inspect screen(s) and/or filter, and leak check the engine. (Level 3)
    g. pre-oil an engine. (Level 2)

L. IGNITION AND STARTING SYSTEMS

*Core competency element

REFERENCE: AP.

Objective. To determine that the applicant:

1. Exhibits knowledge of at least two of the following—

    a. troubleshooting a reciprocating and/or turbine engine ignition system.
    b. replacement of an exciter box and safety concerns if the box is damaged.
    c. troubleshooting a starter system.
    d. checking a starter system for proper operation.
    e. the operation of a pneumatic starting system.
    f. reasons for the starter dropout function of a starter generator or pneumatic starter.
    g. the purpose of a shear section in a starter output shaft.
    h. purpose of checking a p-lead for proper ground.
    i. inspection and servicing of an igniter and/or spark plug.
    j. magneto systems, components, and operation.
    k. function/operation of a magneto switch and p-lead circuit.
    l. high and low tension ignition systems.

2. *Demonstrates the ability to perform at least one of the following (Level 3)—

    a. check engine timing.
    b. check a magneto switch for proper operation.
    c. inspect a turbine engine ignition system for proper installation.
    d. inspect a starter/generator for proper installation.
    e. inspect magneto points.

3. Demonstrates the ability to perform at least one of the following—

    a. install a magneto, and set timing on an aircraft engine. (Level 3)
    b. repair an engine ignition and/or starter system. (Level 3)
    c. remove, inspect, and install turbine engine igniter plugs, and perform a functional check of the igniter system. (Level 3)
    d. inspect generator or starter-generator brushes. (Level 3)
    e. install brushes in a starter or starter-generator. (Level 3)
    f. install breaker points in a magneto and internally time the magneto. (Level 3)

g. repair an engine direct drive electric starter. (Level 3)
h. inspect and test an ignition harness with a high tension lead tester. (Level 3)
i. inspect and/or service and install aircraft spark plugs. (Level 3)
j. bench test an ignition system component. (Level 2)

M. FUEL METERING SYSTEMS

REFERENCE: AP.

Objective. To determine that the applicant:

1. Exhibits knowledge of at least two of the following—

a. troubleshooting an engine that indicates high exhaust gas temperature (EGT) for a particular engine pressure ratio (EPR).
b. purpose of an acceleration check after a trim check.
c. reasons an engine would require a trim check.
d. purpose of the part power stop on some engines when accomplishing engine trim procedure.
e. procedure required to adjust (trim) a fuel control unit (FCU).
f. possible reasons for fuel running out of a carburetor throttle body.
g. indications that would result if the mixture is improperly adjusted.
h. procedure for checking idle mixture on a reciprocating engine.
i. possible causes for poor engine acceleration, engine backfiring or missing when the throttle is advanced.
j. types and operation of various fuel metering systems.
k. fuel metering system components.

2. N/A

3. Demonstrates the ability to perform at least one of the following—

a. remove and install the accelerating pump in a float-type carburetor. (Level 3)
b. check and adjust the float level of a float-type carburetor. (Level 3)
c. check the needle and seat in a float-type carburetor for proper operation. (Level 2)
d. check a fuel injection nozzle for proper spray pattern, and install a fuel injector nozzle. (Level 2)

e.   check and adjust idle mixture. (Level 3)
f.   install a turbine engine fuel nozzle. (Level 3)
g.   locate and identify various fuel metering system
     components. (Level 2)
h.   service a carburetor fuel screen. (Level 3)

N. ENGINE FUEL SYSTEMS

*Core competency element

REFERENCES: AP; JSPT.

Objective. To determine that the applicant:

1.   Exhibits knowledge of at least two of the following—

     a.   inspection requirements for an engine fuel system.
     b.   checks of fuel systems to verify proper operation.
     c.   troubleshooting an engine fuel system.
     d.   procedure for inspection of an engine driven fuel pump for
          leaks and security.
     e.   function and/or operation of one or more types of fuel
          pumps.
     f.   function and/or operation of one or more types of fuel
          valves.
     g.   function and/or operation of engine fuel filters.

2.   *Demonstrates the ability to perform at least one of the following
     (Level 3)—

     a. check a fuel selector valve for proper operation.
     b. inspect an engine fuel filter assembly for leaks.
     c. inspect a repair to an engine fuel system.

3.   Demonstrates the ability to perform at least one of the
     following—

     a.   check a fuel boost pump for proper operation. (Level 3)
     b.   repair fuel selector valve. (Level 3)
     c.   inspect a main fuel filter assembly for leaks. (Level 3)
     d.   check the operation of a remotely located fuel valve.
          (Level 3)
     e.   locate and identify a turbine engine fuel heater. (Level 2)
     f.   service an engine fuel strainer. (Level 3)
     g.   inspect an engine driven fuel pump for leaks and security,
          and perform an engine fuel pressure check. (Level 3)
     h.   repair an engine fuel system or system component.
          (Level 3)
     i.   troubleshoot a fuel pressure system. (Level 3)

O. INDUCTION AND ENGINE AIRFLOW SYSTEMS

*Core competency element

REFERENCES: JSPT; AP.

Objective. To determine that the applicant:

1. Exhibits knowledge of at least two of the following—

   a. inspection procedures for engine ice control systems and/or carburetor air intake and induction manifolds.
   b. operation of an alternate air valve, both automatic and manual heat systems.
   c. troubleshooting ice control systems.
   d. explain how a carburetor heat system operates and the procedure to verify proper operation.
   e. effect(s) on an aircraft engine if the carburetor heat control is improperly adjusted.
   f. causes and effects of induction system ice.
   g. function and operation of one or more types of supercharging systems and components.

2. *Demonstrates the ability to perform inspection of engine induction or airflow system to include at least one of the following (Level 3)—

   a. engine ice control system.
   b. induction manifolds.

3. Demonstrates the ability to perform at least one of the following—

   a. repair a defective condition in a carburetor heat box. (Level 3)
   b. check proper operation of an engine anti-ice system. (Level 3)
   c. rig a carburetor heat box. (Level 3)
   d. inspect an induction system. (Level 3)
   e. replace an induction system manifold gasket and/or induction tube. (Level 3)
   f. service an induction system air filter. (Level 3)
   g. trouble shoot an engine malfunction resulting from a defective induction or supercharging system. (Level 3)

P. ENGINE COOLING SYSTEMS

REFERENCES: AC 65-12A; AP.

Objective. To determine that the applicant:

1. Exhibits knowledge of at least two of the following—

    a. required inspection on an engine cooling system.
    b. operation of cowl flaps, and how cooling is accomplished.
    c. how turbine engine cooling is accomplished.
    d. cooling of engine bearings and other parts on turbine engines.
    e. the importance of proper engine baffle and seal installation.
    f. the operation of a heat exchanger.
    g. the function and operation of an augmentor cooling system.
    h. rotorcraft engine cooling systems.

2. N/A

3. Demonstrate the ability to perform at least one of the following—

    a. inspect an engine cooling system. (Level 3)
    b. check cowl flap operation and inspect rigging. (Level 3)
    c. repair one or more cylinder cooling fins. (Level 3)
    d. repair an engine pressure baffle plate. (Level 3)
    e. inspect a heat exchanger. (Level 3)
    f. troubleshoot an engine cooling system. (Level 3)
    g. locate and identify rotorcraft cooling system components. (Level 2)

FAA-S-8081-28

Q. ENGINE EXHAUST AND REVERSER SYSTEMS

*Core competency element

REFERENCES: AC 43.13-1B, AC 65-12A; AGTP.

Objective. To determine that the applicant:

1. Exhibits knowledge of at least two of the following—

    a. exhaust leak indications and/or methods of detection.
    b. thrust reverser system operation and components.
    c. differences between a cascade and a mechanical blockage door thrust reverser.
    d. hazards of exhaust system failure.
    e. effects of using improper materials to mark on exhaust system components.
    f. function and operation of various exhaust system components.

2. *Demonstrates the ability to perform inspection of engine exhaust system and/or turbocharger system. (Level 3)

3. Demonstrates the ability to perform at least one of the following—

    a. determine if components of an exhaust system are serviceable. (Level 2)
    b. show the procedures to accomplish a pressurization check of an exhaust system. (Level 2)
    c. repair one or more exhaust system components. (Level 3)
    d. check engine exhaust system for proper operation. (Level 3)
    e. replace one or more exhaust gaskets. (Level 3)
    f. install an engine exhaust system. (Level 3)
    g. check a turbocharger and waste gate system for proper operation. (Level 3)
    h. troubleshoot and/or repair a turbine engine thrust reverser system and/or system component(s). (Level 3)

R. PROPELLERS

*Core competency element

REFERENCES: AP; AC 43.13-1B.

Objective. To determine that the applicant:

 1.   Exhibits knowledge of at least two of the following— |

      a.   propeller theory of operation.
      b.   checks necessary to verify proper operation of propeller systems.
      c.   procedures for proper application of propeller lubricants.
      d.   installation or removal of a propeller.
      e.   measurement of blade angle with a propeller protractor.
      f.   repairs classified as major repairs on an aluminum propeller.
      g.   reference data for reducing the diameter of a type certificated propeller.
      h.   operation of propeller system component(s).
      i.   propeller governor components and operation.
      j.   theory and operation of various types of constant speed propellers.
      k.   function and operation of propeller synchronizing systems.
      l.   function and operation of propeller ice control systems.

  2.   *Demonstrates the ability to perform both of the following—

      a.   inspection of a propeller installation, and make a minor repair on an aluminum propeller. (Level 3)
      b.   determine what minor propeller alterations are acceptable using the appropriate type certificate data sheet. (Level 2)

  3.   Demonstrates the ability to perform at least one of the following—

      a.   service a constant speed propeller with lubricant. (Level 2)
      b.   use a propeller protractor to determine correct blade angle. (Level 3)
      c.   leak check a constant speed propeller installation. (Level 3)
      d.   install a fixed pitch propeller and check the tip tracking. (Level 3)
      e.   inspect a spinner/ bulkhead for defects and proper alignment and installation. (Level 3)

      f.   dye-penetrant inspection to determine the amount of propeller damage. (Level 2)

      g.   inspect and/or adjust a propeller governor. (Level 3)

      h.   inspect a wood propeller. (Level 3)

      i.   troubleshoot a propeller system. (Level 3)

S. TURBINE POWERED AUXILIARY POWER UNITS

REFERENCE: AP.

Objective. To determine that the applicant:

1. Exhibits knowledge of at least two of the following:

      a.   inspection to ensure proper operation of turbine driven auxiliary power unit.

      b.   replacement procedure for an igniter plug.

      c.   servicing an auxiliary power unit.

      d.   troubleshooting an auxiliary power unit.

      e.   function and operation of auxiliary power unit(s).

NOTE: Subject area T, AUXILIARY POWER UNITS, may be tested at the same time as AREA B, TURBINE ENGINES. No further testing of auxiliary power units is required.